Facilities Change Management

D1579720

Facilities Change Management

Edited by

Edward Finch
Professor in Facilities Management
The University of Salford

WILEY-BLACKWELL

A John Wiley & Sons, Ltd., Publication

This edition first published 2012 by Blackwell Publishing Ltd
© 2012 Blackwell Publishing Ltd

Wiley-Blackwell is an imprint of John Wiley & Sons, formed by the merger of Wiley's global Scientific,
Technical and Medical business with Blackwell Publishing.

Registered office:
John Wiley & Sons, Ltd, The Atrium, Southern Gate, Chichester, West Sussex, PO19 8SQ, UK

Editorial offices:
9600 Garsington Road, Oxford, OX4 2DQ, UK
The Atrium, Southern Gate, Chichester, West Sussex, PO19 8SQ, UK
2121 State Avenue, Ames, Iowa 50014-8300, USA

For details of our global editorial offices, for customer services and for information about how to
apply for permission to reuse the copyright material in this book please see our website at
www.wiley.com/wiley-blackwell.

Library of Congress Cataloging-in-Publication Data

Finch, Edward.
  Facilities change management / edited by Edward Finch.
    p. cm.
  Includes bibliographical references and index.
  ISBN-13: 978-1-4051-5346-1 (pbk. : alk. paper)
  ISBN-10: 1-4051-5346-6
 1. Buildings–Remodeling for other use. 2. Facility management. I. Title.
  TH3411.F46 2011
  658.2–dc23   2011030350

A catalogue record for this book is available from the British Library.

This book is published in the following electronic formats: ePDF 9781444346060; Wiley Online Library
9781119967316; ePub 9781444346077; MobiPocket 9781444346084

Set in 10/12pt Minion by Thomson Digital, Noida, India.
Printed and bound in Malaysia by Vivar Printing Sdn Bhd

1 2012

Contents

Preface

FACILITIES CHANGE MANAGEMENT

It would appear that we no longer need many buildings — or so some would have us believe. In their view, offices, hospitals, prisons and education buildings, among others, are becoming a relic of the past. That is the argument put forward by the advocates of virtualisation, sometimes on the grounds of carbon reduction, convenience or cost savings. Technology, it is suggested, is brushing aside the need for such costly physical assets. Home working, community based health, electronic tagging and online learning are presented as inevitable alternatives that make the need for such building types redundant.

The evidence however suggests otherwise. The demands for such buildings are just as great as ever, despite the apparent option of remote working, remote telemetry and remote learning. How can this be explained? People are sentient beings, who seek the stimulation that the built environment presents. They embrace the opportunity to be part of a vibrant physical congregation, an organisation, housed in a facility with a definable purpose. What is evident is that the future demand for such buildings is assured, in meeting the needs and motivations of individuals. However, just what form they take is much less certain. Businesses, public authorities and local communities have to rethink exactly what is required by the familiar concepts of school, police station, hospital or workplace. In short, they require reinvention. The design of such buildings is no longer tethered by constraints of the past. New technology satisfies the need to access resources that previously could only be met by a journey to this building or that one. Now that such a need can be met in other ways, we are forced to question the very rationale for their existence.

This book in facilities change management is designed for those entrusted with this challenge — the challenge of making the physical environments we inhabit fit for a future that will be significantly different. Such professionals include facilities managers, property managers, architects, building users and those responsible for investing in our future environments.

Looking specifically at the role of the facilities manager, which previously had been described as a 'Cinderella' profession in the construction industry, it is now being challenged with deep, searching questions about how to meet the demands of future building users. This professional stance is unfamiliar to many facilities managers, whose track record has traditionally been proven in terms of 'delivery' — can you deliver on time and to budget with the least amount of aggravation to building inhabitants? In short, can you deliver what the client thinks he needs?

This unquestioning mental position, inevitably leads to a process of self talk of the form 'we can achieve anything' and 'nothing can stop us'. This attempt to banish doubts is seen to be part of the recipe of the facilities manager's success. However, it may indeed be their undoing. Enabling a degree of doubt to enter into the equation might be what is required.

A recent study by three US social scientists (Ibrahim Senay and Dolores Albarracin of the University of Illinois, along with Kenji Noguchi of the University of Southern Mississippi) explored the difference between what is called 'declarative' self talk ('We will get it done') to 'interrogative' self-talk ('Can we get this right?'). A simple experiment involving the resolution of some anagrams was conducted. But prior to this, participants were split into two groups: one half took a minute to consider whether the task could be completed; the other half took the time to tell themselves that they could complete the task. The result? The group of self-questioners was able to resolve significantly more anagrams than the self-affirming group. So what is going on here? One of the researchers, Albarracin, explains in a UK national newspaper (cited by Pink, 2011) that the process of '. . . setting goals and striving to achieve them assumes, by definition, that there is a discrepancy between where you are and want to be. When you doubt, you probably achieve the right mindset'.

The authors in this book raise just such seeds of doubt in relation to the management of change in the built environment. They make no apologies for doing this and common to each of the chapters is a searching and questioning predisposition. Leading authors from Australia, Brazil, Canada, Netherlands, Turkey, China and the UK, among others, each present part of a holistic framework for raising such questions and developing the evidence base to resolve them. It is hoped that this book will be instrumental in supporting a new generation of facilities professionals that are able to ask the right questions.

Pink, D.H. (2011). 'Can we fix it' is the right question to ask. *The Daily Telegraph*, 29 May.

Edward Finch, Editor

Contributors

Cláudia Miranda de Andrade
Andrade Azevedo Arquitetura
Corporativa
São Paulo, Brazil

Simon Austin
Department of Civil & Building
Engineering
Loughborough University
Loughborough
UK

Iris de Been
Center for People and Buildings
Delft
The Netherlands

Melanie Bull
Sheffield Business School
Sheffield Hallam University
Sheffield, UK

Tim Brown
Sheffield Business School
Sheffield Hallam University
Sheffield, UK

Paul Dettwiler
Institute of Facility Management
Zurich University of Applied Sciences
Zurich, Switzerland

Edward Finch
School of the Built Environment
University of Salford, UK

Alistair Gibb
Department of Civil & Building
Engineering
Loughborough University
Loughborough, UK

John Hudson
School of the Built Environment
University of Salford
Salford, UK

Goksenin Inalhan
Faculty of Architecture
Istanbul Technical University
Istanbul, Turkey

Peter Love
School of Built Environment
Curtin University
Western Australia, Australia

Maartje Maarleveld
Center for People and Buildings
Delft, The Netherlands

Sheila Walbe Ornstein
Faculty of Architecture and Urbanism
University of São Paulo
São Paulo, Brazil

James Pinder
Department of Civil & Building
Engineering
Loughborough University
Loughborough, UK

Ana Pereira Roders
Faculty of Architecture, Building and
Planning
Eindhoven University of Technology
Eindhoven, The Netherlands

Rob Schmidt, III
Department of Civil & Building
Engineering
Loughborough University
Loughborough, UK

Jim Smith
Institute of Sustainable Development &
Architecture
Bond University
Queensland, Australia

Danny Shiem Shin Then
Hong Kong Polytechnic University
China

Jacqueline C. Vischer
School of Industrial Design
University of Montreal
Canada

Theo J.M. van der Voordt
Department of Real Estate & Housing
Faculty of Architecture
Delft University of Technology
Delft, The Netherlands

1   Facilities Change Management in Context

Edward Finch

CHAPTER OVERVIEW

The number of books, training seminars and missives on the subject of change management continues to grow unabated. Yet few of these consider the importance of the *physical* change which inevitably accompanies the change of 'minds'. It is the physical change in the form of workplace redesigns, procurement of new buildings or perhaps the reengineering of a facilities service, which present the tangible evidence of change. People often discard the wise words which appear in the mission statement or the new process hardwired into the corporate intranet. If change is going to succeed, evidence suggests that a transformation in what we see, touch and experience is the only kind of change that people within an organisation are likely to understand and internalise.

How does the facilities manager achieve such transformations? A starting point in this journey is the process of 'sense making' or understanding the nature of change. This chapter describes the changing landscape in which facilities management teams operate. In so doing, it seeks to contextualise facilities management. This chapter explains how each of the elements of the change management process is addressed in each of the book's chapters. This is achieved by (1) an analysis of current thinking on change management; (2) an exposition of how facilities management needs to be redefined to accommodate contemporary approaches and (3) an explanation of a framework (described as the REACTT model) which identifies the key stages of facilities change management which in turn correspond with each of the chapters of this book.

Keywords: Change context; REACT model; Facilities management definition; Punctuated change; Transformation.

1.1   FORCES OF CHANGE AFFECTING THE BUILT ENVIRONMENT

A change can be described as any 'alteration in the state or quality of anything' (Shorter English Dictionary). Changes can involve people, technology, services or buildings. Indeed, most changes of any significance impact on a number of these facets. Thus, the facilities manager is never entirely concerned with buildings in isolation. One of the most popular quotes in the field of architecture is that of Winston Churchill:

> We shape our buildings; thereafter they shape us. (Winston Churchill, 1874–1965)

This prophetic observation is indispensable to our understanding of facilities in the context of organisations. It makes clear that the buildings which we find ourselves in are at the outset an expression of all the elements that go to make up an organisation. They represent an expression of its people, what they stand for, their mode of operation, as well as their actual and espoused values. The quote highlights that from the day a facility is occupied such buildings themselves become the agents of change (or inertia). A modern day counterpart to this quote given by Denison and Mishra (1995) contends that:

> Structures are both the medium and the outcome of interaction. They are the medium because structures provide the rules and resources individuals must draw upon to interact meaningfully. They are its outcome, because rules exist only through being applied and acknowledged in interaction – they have no reality independent of the social practices they constitute. (Denison and Mishra (1995), p. 206)

Ironically, the quote conceives of structures as organisational structures. However, it is most apt in describing the importance of 'physical structures' (buildings, floor layouts and supporting services) which provide the 'hardwired' rules that dictate organisational interaction and social practices.

1.2   INERTIA AND CHANGE

Early thinkers on the nature of change construed change as an incremental process. This view of the world is described as the 'gradualist' paradigm. Continuous improvement (*Kaizen*; Japanese for 'change for the better') was proposed as the key method for managing change in an environment which was perceived as largely predictable. Based on this concept, changes to individual subsystems such as people, missions or facilities provide the necessary intervention to allow small but continuous change that allows adaptation to the internal and external environment. In such a model it is possible to tinker with one part of the system without affecting the whole (Choi, 1995).

However, in parallel with modern day reinterpretations of biological evolution, it has been argued by Gersick (1991) that change in most organisations is not continuous, but is characterised by events involving rapid change. In just the same way as evolution in the natural world undergoes major transformative events, so it can be seen that organisations are also subject to such rapid and often unexpected change. Gersick (1991) studied change in individuals, groups and in organisations as well as the history of science. She found in all of these change categories a recurring pattern of relatively long periods of stability or equilibrium 'punctuated' by short bursts of metamorphosis. The paradigm known as 'punctuated equilibrium' was used to describe this pattern.

How can we explain this process of punctuated equilibrium and more importantly what are the ramifications for facilities management? The model is explained in terms of in-built organisational inertia which arises from persistent *deep structures* which allow only small incremental changes. It is these embedded structures which resist change and pull an organisation back to a condition of equilibrium. Such deep structures are highly stable. This stability arises from the establishment of a number of key choices in the organisation's history that exclude many options which might be deemed inconsistent. These mutually interdependent choices reinforce and strengthen one another over time. Gersick (1991) suggests that three sources of inertia are at play in organisations.

- Sense making: the organisation's way of seeing things (cognitive framework). Organisations evolve shared mental models in the way that they interpret reality and learn. In reaction to change, the natural response is to look at ways of 'doing things better'. Notice that this contrasts with a more open approach which considers all options and also considers doing 'better things'. The focus is thus on efficiency and alignment rather than the exploration of new opportunities.
- Motivation: change brings with it a fear of loss as well as a realisation that such change may bring about a 'sunk cost'. For example, the change in choice of air-conditioning manufacturer may render the expertise of a plant engineer redundant, having gained years of experience in the maintenance and regulation of an existing system.
- Obligation: with any change comes disruption and the severing of interdependencies. Relationships with particular service providers may have to be terminated: short-term disruptions to customer services may ensue. In the short term, the attraction of change may be lacking and the turmoil and loss of goodwill may be the dominant concern.

At some point in time, the forces of inertia, despite their attractive forces (e.g. efficiencies achieved through interdependency) become overwhelmed by external changes. The ensuing change is inevitably shattering to the status quo, resulting in 'punctuation in time'.

The punctuated equilibrium paradigm explains much about why facilities management change initiatives are so often challenging and problematic. Initiatives such as hot-desking or energy-awareness are often only able to impact on the outer superficial structures without penetrating more resilient deeper structures (culture and behaviour). This explains the legacy of failed attempts to introduce 'new ways of working' that directly challenge the deep structures of an organisation. Much of the literature on change in facilities is founded on the 'gradualist' paradigm of change, whereby, through a process of continuous adjustment, it is possible to respond to the changing environment.

As well as explaining why facilities managers encounter resistance when implementing change, the punctuated equilibrium model also highlights the significance of major facilities initiatives. The decision to relocate from a central business district to a suburban location may coincide with a change in business model, for example, a change from face-to-face to online customer service support. The relocation is thus a 'punctuation' or radical departure from the past way of doing things: the facilities change is simply a physical manifestation of a deep structural change. As such, every opportunity is taken in such a move to realign and transform systems: not as separate systems but as part of a holistic entity. The opportunity for transformative change which arises from a relocation or change in service provider is clear. However, it is incumbent on the facilities manager to realise this opportunity. As such, it involves working in concert with other systems within an organisation to overcome deep structural inertia.

1.3   UNDERSTANDING THE S-CURVE

Facilities management is driven by space forecasts and space budgets. Often such forecasts of change are based on simple extrapolations of what has gone before. A pattern of 5% growth in personnel over the last ten years is assumed to be repeated for the next five years.

However, such a forecasting approach is fraught with dangers based on business as usual. One of the most widely recognised predictor of change, the sigmoidal curve or S-curve, has consistently proven to be a reliable tool. Early work by Tushman and Romanellli (1985) showed how the S-curve (so described because of its characteristic S-shape) accurately describes the growth pattern of innovations and organisations. The curve illustrates the slow growth rate associated with start-up organisations whose initial growth is tempered by resource constraints and market acceptance. This is then succeeded by a period of rapid (exponential) growth during which time the organisation undergoes successive periods of growth. Finally, as the service or product offering is exhausted, the growth rate reaches maturity, with a tapering of growth.

The S-curve equally describes the characteristic growth patterns of individuals and indeed their lifestyles. This was illustrated in the seminal work of Becker (1990) in *The Total Workplace*. This describes how life changes impact on our choice of residential property. This begins with modest requirements to support a singles lifestyle, succeeded by shared living and the arrival of a family. Finally, in the maturing stage, married couples become 'empty nesters' and begin to downsize. At each stage there is punctuation in their existence associated with a change of accommodation. Houses and apartments are bought and sold to realign with their changing needs.

Figure 1.1 illustrates organisational growth in the form of an S-curve and the way this impacts on facilities and relocation decisions. During the early stages of growth, organisations typically occupy 'incubator' facilities, providing the flexibility for experimentation unencumbered by constraints and standards. The organisation at this stage is involved in inventing its 'deep structure'. As the organisation becomes too large for its original facility, the pressures for a transformative change overwhelm the forces of inertia. Thus a 'punctuation' in an organisation's timeline occurs. The organisation has established its deep structures, including its characteristic way of doing things, its mission and underlying

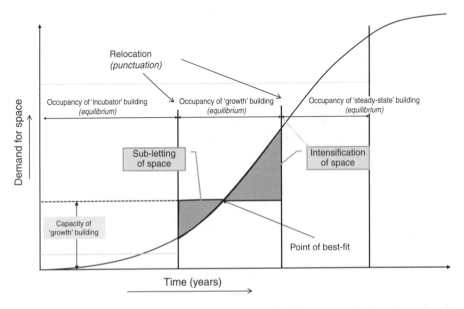

Figure 1.1 S-curve patterns of growth and space demand within an organisation. Reproduced by permission of John Wiley & Sons, Inc.

culture. The new building (unlike the incubator building) attempts to formalise and express this emergent view of itself. At the same time, the organisation seeks out standards and efficiency measures. This is what is described by Becker (1990) as the transition from the 'loose fit' to the 'tight fit' organisation. This in turn is reflected in the facilities management operation, with the emergence of formal policies and standards (such as space standards) which attempt to rationalise the service provision. Becker also refers to the concept of 'elastic fit' as the form of building solution (and facilities management solution) which succeeds the 'tight fit' approach. The constraints imposed by operating standards, whilst allowing efficiencies, prevent the organisation from growing. Only by engineering a degree of flexibility can an organisation extend its life on the S-curve: it needs to reinvent itself. The 'elastic' model relinquishes formal 'standards' in favour of 'frameworks' and 'templates' tuned to the individual needs of each part of the organisation. During the growth and downsizing stages, further moves and relocations inevitably occur, ranging from the relocation of individuals or departments to whole organisations. Moves inevitably are associated with 'punctuation' — an opportunity for transformation. They present a chance to realign the space and service offering; an opportunity to bridge the emerging gap between what is required and what is available. Treating the relocation as simply a resizing operation is fundamentally flawed. Such a 'punctuation' needs to embrace all of the levels of change in an organisation; in other words, to enable a 'punctuation' which addresses some or all of the forces of inertia.

Figure 1.1 also illustrates the constant state of inexact fit which arises between the space demands of the organisation (the demand side) and the capacity of the building (the supply side). At almost no point is entire equilibrium between these two forces met. The shaded zones identify areas where the facilities manager is constantly having to compensate for the deficiencies in this mismatch. During the early period of occupancy (or during periods of downsizing) there may be a surfeit of space. Facilities managers, however, are rarely presented with this problem as departments undertake unsanctioned 'creep' into unoccupied areas. The problem then becomes one of preventing encroachment and of using excess space. Sub-letting is one such approach, as is the homogenising of capacity between more than one building. However, having too much space can be as challenging as having too little space. Beyond a specific 'tipping point' in the organisation's growth cycle, the challenge becomes one of limited space. Measures to intensify the use of space (in tandem with innovations such as hot-desking or desk-sharing) can significantly extend the capacity of the building. Indeed, more and more buildings are built on this premise from the outset. Even for large multinational organisations, this problem of fit in relation to a single building or campus remains. The option of distributing staff between geographical locations may not be an option where co-location or adjacency to market is an imperative. The S-curve becomes an essential weapon in the facilities manager's armoury in forestalling simplistic projections.

1.4   THE CONTEXT OF CHANGE

Most theory relating to change management lacks context. Many address only an isolated aspect of change (e.g. culture, workplace design, process design); whilst others put forward a single approach to change (e.g. 'organisation development', 'systems thinking' or 'strategic planning'). No wonder the facilities management profession is reluctant to spend time making sense of such a convoluted collection of 'recipes'. However, as we have seen in the previous section, the central role of facilities in the 'punctuated equilibrium' paradigm makes an understanding of context essential. Without it, facilities managers cannot hope to use the serendipity of the move process to realign with 'deeper structures'.

Mintzberg and Westley (1992) are one of very few authors who acknowledge the central role played by facilities and supporting services in realising change. The authors attempted to confront the problem wherein 'by seeking to explain the part, we distort the whole'. In their key publication 'Cycles of Organizational Change' they describe a 'system of moving circles' (like a bulls-eye) to represent the various aspects of change at differing levels of abstraction.

Using their contextual model of the organisation, change is seen as taking place from the broadest conceptual level (i.e. in the minds of organisational thinkers) to the most concrete and tangible level (facilities and people). But, they argue, such change also occurs in one of two spheres; (1) the basic *state* and (2) the thrust or *direction* of the organisation. Put together, we end up with the total landscape of change which confronts an organisation.

1.4.1 State versus direction

The two facets of 'state' and 'direction' define the two spheres of activity facing the modern-day facilities manager, as shown in Table 1.1. The 'state' is about what you have got. We can reconfigure organisational services, delivery systems and people to satisfy the changing requirements of an organisation. Service level agreements can be modified, maintenance staff can be redeployed and space plans rearranged. Such changes often occur incrementally, over days, months or years, often in a piecemeal manner (what Mintzberg calls the 'deductive' change). Whilst these changes may be largely unplanned, over time they can have a profound effect on an organisation. Contrast this with the second facet, 'change in strategy' which involves planned change and determines the direction of an organisation. It is this second sphere which is increasingly the territory of the facilities manager, demanding a project management approach and longer term planning.

It is these two distinct facets which exemplify the two different worlds of 'facilities management' and 'real estate'. The two were often set apart, with facilities solutions allowed to *evolve* in response to changing demands. In contrast, real estate involved key strategic commitments with the impacts unfolding over many years, often as part of a masterplan. However, things have changed fundamentally. New innovations in workplace design, sustainable transport planning and energy initiatives, among others, have transplanted the activities of the facilities manager from evolutionary planning to project planning. The worlds of facilities management and real estate have begun to converge. Real estate has started to adopt the language of 'user-centred' behaviour (state), whilst facilities managers have embraced visionary language (strategy) previously confined to the real estate role.

Table 1.1 Elements of change (adapted from Mintzberg and Westley, 1992).

	Change in organization: State	Change in strategy: Direction
Thinking level (abstract)	culture organisational structure	vision positions
Doing level (concrete)	systems facilities services people	programs facilities assets

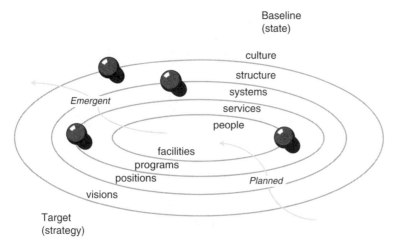

Baseline
(state)

culture
structure
systems
services
people

Emergent

facilities
programs
positions
visions

Planned

Target
(strategy)

Figure 1.2 Orbital view of change (adapted from Mintzberg and Westley, 1992).

In Figure 1.2 we present the 'orbital model' of change management based on the *moving cycles* concept of Mintzberg and Westley (1992). Like the original model, it recognises that change can take the form of radial change (moving from the outer abstract level to the inner concrete level). However, unlike the spiralling view of Mintzberg and Westley, the orbital model proposes that a discrete quantum of 'energy' needs to be expended in order to move up through to outer levels. Without the necessary energy input, the change management initiative falls back to its original state. This energy might be referred to as the 'transition energy' (using the parlance of atomic theory). Change can also occur at a single orbital level (what Mintzberg and Westley calls 'circumferential' change) so that the impact of facilities changes may be confined to people and workplace, without a more abstract impact on structure or culture within the organisation. This gives rise to isolated pockets of change, either at a discrete point on the orbital path (isolated change) or within a single ring (silo change). Thus, the introduction of 'new ways of working' may be introduced at the 'people' or 'facilities' level; but the required transformation is unlikely to take place without the additional input of energy to affect the 'systems', 'structure' or 'culture' level. At the other extreme, an orbiting object (change initiative) may reach 'escape velocity', propelling the organisation into a new 'realm' of sequences and patterns of change.

So who needs another change management model? The value of Mintzberg and Westley's original model is that it places facilities management 'centre stage', unlike most change management models which make only passing reference to facilities as a vehicle of change. For those seeking to leverage support for facilities management initiatives, the ability to 'contextualise' the change is paramount. Is it an isolated approach, a piecemeal approach or perhaps a revolutionary approach? The orbital model highlights those areas of change where further investment of energy is required to transform an isolated or piecemeal approach into a more focused or revolutionary approach. The model demonstrates graphically the two realms of 'real estate management' and 'facilities management'. Whilst the former is typically driven by formal change, often in the form of a structured acquisition process or perhaps as part of a masterplan (the top-down approach), facilities management makes much more use of a bottom-up approach, in the form of informal, emergent change.

For corporate organisations seeking to roll-out change on a global arena, having a holistic view of change is an overriding concern. Such an approach allows innovation to be fully exploited in various geographical regions and at different levels of abstraction within an organisation. The proposed 'roadmap' provides a useful tool that will allow more effective engagement between the 'concrete' world of the facilities management team and the 'conceptual' aspirations of organisational leaders.

1.5 FACILITIES MANAGEMENT AND THE BUSINESS OF CHANGE

So far in our discussion we have highlighted the emergence of the facilities manager as 'project manager'. But what do we understand by the term 'facilities management' and, more importantly, what is the reality in practice? The definition of 'facility management' provided by the European Committee for Standardisation (CEN) EN 15221-1 is given as:

> the *integration* of processes within an organisation to maintain and develop the agreed services which support and improve the effectiveness of its primary activities.

Such a definition, whilst describing the scope of facilities management, does not attempt to provide an explanation regarding the skills or competencies required to undertake this task. There is no mention of innovation, leadership or skills development. Instead, the definition concerns itself with systems and system boundaries between contracting parties. What are primary activities? As noted by Barrett and Baldry (2003), facilities and their supporting services may themselves be the primary activity (e.g. hotels) or may secondarily become part of the primary activity (e.g. hospital cleaning as part of patient care).

More problematic with the CEN/definition are the three levels that are assumed to exist: (1) strategic; (2) tactical and (3) operational. The 'operational' level appears to underpin the facilities management activity. Quite at odds with this view, we see increasing evidence that facilities managers are not operational managers but are indeed project managers. They are involved in transformations, refurbishment projects, remodelling or relocations that have a discrete 'start' and a discrete 'end'. Such change management projects draw on skills which are not encompassed by the term 'integration' or indeed 'operations'. The focus is very much on the 'develop' rather than 'maintain' philosophy.

Section A.2 of the standard acknowledges the role of facilities management in the change process:

> An organisation relies on its primary processes in order to achieve its strategic objectives. Changing market forces and developments from legislation, technology and mergers constantly influence these processes. These changes shall be managed and structured in strategic, tactical and operational levels, in order to remain viable and compliant to changing demands. (EN 15221-1 (CEN/TC348, 2006))

However, what this statement does not recognise is the transformative potential of facilities and facilities management. The profession is concerned with much more than 'remaining viable' (efficiency) and being 'compliant'. Returning to our earlier observation, buildings and facilities themselves have the potential to shape and change the organisation. They are not simply an instrument of change — they *are* the change.

The European standard EN 15221-1:2000 describes a *systems* (process) perspective of facility management operating at the strategic, tactical and operational level. This assumes a set of interrelated components that transacts with the larger environment. Interestingly, the definition makes no reference to the role of the built environment. In the context of this book, the physical environment is identified as playing a central role – that processes alone do not define facilities management. Rather, it is the capacity of buildings and building systems to guide 'human behaviour'. As such, the challenge of facilities management in the context of change becomes much greater. Instead of being *equilibrium* seeking, based on a steady state strategy, facilities management recognises the importance of punctuated equilibrium in a non-steady state environment.

The nursing profession has an underlying set of principles that are equally applicable to the facilities management profession. The *International Council of Nurses* define the role of nursing as encompassing:

... autonomous and collaborative care of individuals of all ages, families, groups and communities, sick or well and in all settings. Nursing includes the promotion of health, prevention of illness, and the care of ill, disabled and dying people. Advocacy, promotion of a safe environment, research, participation in shaping health policy and in patient and health systems management, and education are also key nursing roles. (International Council of Nurses (2010))

By minor modifications, it is possible to identify an equally fitting definition of facilities management:

Facility management encompasses in-house and collaborative provision of service settings for individuals, organisations and communities. Facility management enables the promotion of organisational effectiveness and individual wellbeing by leveraging the transformative potential of such service settings. Also key to the facilities management role are advocacy in shaping organisational policy, promotion of a healthy environment, research and professional development.

This definition serves a different purpose to that envisaged in the European standard. Whilst the European standard attempts to demarcate the industry boundaries, the definition above seeks to identify the professional role of the facilities manager. Such a definition is not predicated on a systems view of facilities management; rather, it acknowledges non-steady state change issues (shaping and research) and it is clearly tied to space and the built environment (service settings).

1.6 THE SCOPE OF FACILITIES CHANGE MANAGEMENT

What supporting services does the facilities manager need to integrate to achieve a suitable service setting? Figure 1.3 describes an organisational structure for the facilities team proposed by Williams (1996). This structure incorporates three key strands:

- premises
- support services
- information technology.

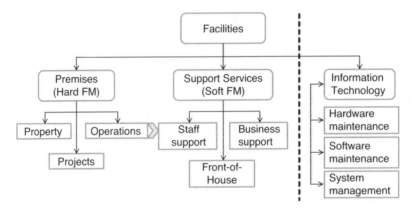

Figure 1.3 The scope of facilities management (adapted from Williams (1996)).

Williams (1996) goes on to suggest that 'the importance of involving the information technology (IT) function under the umbrella must never be underestimated.' He argues that information technology forms an integral part of any change issues related to workplace arrangements, location and facilities requirements.

The premises responsibility is further subdivided into three activities: property, projects and operations. As shown in Figure 1.3, 'property' would typically involve all activities involved in procurement and disposal of property assets and leases, to produce a property portfolio that meets the emerging needs of an organisation. The 'projects' function is clearly aligned with change management agendas, involving strategic decision making and financial investment. This activity also presents challenges in terms of the teams engaged, with the number and constitution of project teams changing over the course of the project. It is through the support of such projects that it is possible to extend the useful life of assets, through adaptation and upgrading. Projects may include simple churn issues related to the movement of staff or departments. They may extend to the design and construction of new buildings or even masterplans. The 'operating' role is in essence concerned with 'day-to-day' steady state activities including the important functions of cleaning, security and maintenance. However, even these seemingly operational activities take on a 'projects' characteristic in many situations. Thus, infrequent external cleaning of facades may have few operational characteristics, requiring a risk assessment and innovative access strategies to clean inaccessible areas such as atria. Similarly, security may not fit the 'operational' tag in all situations. The attendance of VIPs or an unexpected security situation may prompt a change management approach that is far from being operationalised.

Adaptations to the original diagram of Williams (1996) relate to the interface between 'operations' and 'staff support'. The emergence of participative facilities management and the increased engagement of employees in the facilities agenda makes this interface a particularly important one. Figure 1.3 also highlights the increasing importance of 'front-of-house' services in addition to staff and business support. Service settings in this context are used to accommodate visitor needs and to communicate the corporate brand to such visitors. Reception and concierge services represent an important part of the facilities management role in this respect.

1.7   REPLACING LIKE WITH UNLIKE

Rarely do designers now replace like with like, particularly in fast advancing technological areas such as mechanical and electrical services. Such systems inevitably exhibit declining performance as they age. The replacement of these items may occur every 5 years (e.g. diffusers) or every 20 years (e.g. plant and boilers) and in the intervening period regular maintenance can extend their life. In an unlikely world of unchanging needs (legislation, technology and organisational change) it is sufficient to replace like with like to bring the buildings condition up to an 'as new' state of repair. However, such a strategy is unlikely to meet the changing demands that produce an advancing utility gap. In Chapter 3, further consideration is given to the inevitable process of declining performance through age. At the same time, needs change along with changes in legislation and technology. The inevitable decline in performance of plant, services and fabric over time eventually prompt a replacement decision. The facilities manager is thus called upon to do more than 'maintenance' (i.e. merely retaining the 'status quo'). Technologies change such that new performance standards are possible. Legislation and standards also change, so that in the case of boiler and refrigeration plant, new energy and environmental standards dictate a change in product and a change in the way it is operated. Thus we see that 'change management' is integral to the modern facilities manager's role.

1.8   THE INTELLIGENT CLIENT

In order to harness the benefits of change, the typical role of the facilities manager as 'fire fighter' is not sufficient. Many modern organisations are pursuing the objective of being a 'thin client' in relation to facilities management, retaining only a minimal in-house management capability. Pressed with organisational demands and the need to focus on core business, for many organisations, outsourcing presents an attractive proposition. Using the outsourcing approach, management expertise, technical expertise and labour are procured from the external market by means of a contract. Is it possible for an organisation to relinquish all responsibilities using this approach? Williams (1996) suggests that there is a core capability that needs to be retained in-house. This facilities management capability can be described as the 'intelligent client'. Essentially, this provides a senior management capability which is able to represent the interests of the organisation. In order to do this, three facets need to be addressed:

1 sponsorship (policy and strategy), involving:
   a creation (support from within and outside of the organisation)
   b strategy formulation
   c changing
   d directing
2 intelligence (understanding and monitoring)
   a customer objectives
   b customer needs
   c technology
   d service delivery

3 service management (contract management)
  a agency
  b task management
  c contract management

Williams (1996) goes on to argue that 'the proper allocation of resources to each of these facets is absolutely critical to the achievement of cost effective facilities management.' He suggests that many organisations often possess only one or two of these three essential facets (sponsorship, intelligence and service management). Organisations invariably focus on the procurement process (service management role) as an opportunity to 'squeeze out' savings. However, the opportunities for budgetary control (enabled through the forward-looking sponsorship role) and the opportunities for value engineering (enabled through the intelligence role and an understanding of need) are two equally important areas. Both of these facets do not need to come at the expense of the outsourced service provider's profit margin. A mature partnering relationship with an outsourced provider should explore the possibilities for budgetary control and value engineering.

1.9   THE CHANGE MANAGEMENT CYCLE

So far we have identified two characteristics of change that represents the greatest challenge and greatest opportunity in facilities management: (1) the punctuated nature of change, whereby changes to facilities impact on many other dimensions of an organisation, and (2) the focus on human engagement which is central to the eventual success of any change management challenge. To this end, we have noted that facilities managers are not just integrators, they need to be leaders of change. Invariably, such change requires active intervention through the various phases of change. The temptation is to concentrate on the 'doing' part of a change management process (the move, the signing of a service contract or the installation of a new technology), since this is often the most conspicuous activity which often attracts the attention of managers, stakeholders and the press. However, success is largely dependent on careful analysis before the event and further refinement after the event.

This book directly tackles the issues confronted by facilities managers in an environment of change. The sequence in which the chapters appear reflect the cycle of interventions required by the facilities management team. The same sequence applies to both 'physical assets' and 'procured services'. Whilst it is often tempting to assume that the building alone is the subject of change, in reality there are many other smaller change cycles that cumulatively impact on building performance. New technologies, new working practices, reorganisations, new service solutions – all have their own adoption cycle. With this in mind, the phases through which change occurs are not described solely in terms of the whole building's lifecycle. Despite this, we can see parallels with models such as the RIBA (1999) Plan of Works which encompasses (1) Appraisal; (2) Strategic Brief; (3) Outline Proposal; (4) Detailed Proposal; (5) Final Proposal; (6) Production Information; (7) Tender Documents; (8) Tender Action; (9) Mobilisation; (10) Construction to Practical Completion; (11) After Practical Completion. Such a model, however, is designed for the purposes of managing a process rather than challenging assumptions and identifying the need for change.

The chapter sequence used in this book reflects a change management cycle as shown in Figure 1.4. It is based on a modification of the REACT model of Macrimmon and

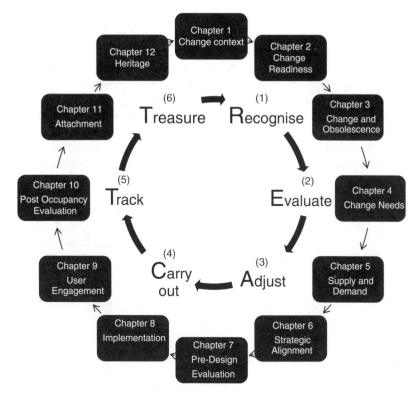

Figure 1.4   REACT model – the framework for 'facilities change management'.

Wehrung (1988), a risk management model that lends itself to an understanding of transformations. In the original model the key stages are (1) recognise; (2) evaluate; (3) assess; (4) choose; (5) track. The modified model (which describes fully the organisation of this book) aligns with the following stages of what is described as the REACTT model:

- Recognise
- Evaluate
- Assess
- Carry out
- Track
- Treasure

The relationship between the book chapters and each of these stages is illustrated in Figure 1.4 and explained further in the following sections.

1.9.1   Recognise

How do we avoid working on the wrong problem? This is what the well-known psychoanalyst Freud called the 'presenting problem'. He observed that when patients were asked what their problems were, their true concerns did not become apparent without further probing. This framing problem is often the most challenging aspect of facilities change management:

recognising that you have a problem in the first place. It is one aspect of the decision-making process which defies attempts to automate a change management process, requiring a uniquely human understanding.

Chapter 1 (Finch) has provided an overview of the change management challenge for the facilities management profession. Putting the problem in context is central to any facilities management undertaking. Invariably, what appears to be a simple mobilisation activity soon emerges as a problem which touches on the soft underbelly of an organisation (deep structures).

Chapter 2 (Finch) addresses the challenge of change readiness. This describes the ability to configure systems and assets such that they can meet the requirements of various possible futures. Again, not only do we have the 'presenting problem' of current day concerns, we also need to identify concerns of tomorrow.

Chapter 3 (Pinder) considers the concept of obsolescence. Obsolescence is an inevitable consequence of the passage of time. The challenge therefore is to anticipate this process and take the necessary steps to prolong the useful life of assets. The chapter identifies two interrelated approaches for coping with changing user demands: designing buildings for adaptability and adaptive re-use. Both approaches rely on the initial recognition of a facility's inherent potential.

1.9.2 Evaluate

Having recognised future challenges, it is then necessary to understand the baseline position. How do the assets that we have inherited and the aspirations of the organisations align with the future challenges 'presented' in the recognition phase?

Chapter 4 (Dettwiler) highlights the requirements for capturing client 'needs' — not based on wishes or desires but on sound evidence. This process is illustrated in relation to start-up organisations whose critical facilities decisions at the growth stage can have a profound effect on the future development of an organisation. By choosing the wrong facilities options, the start-up becomes entangled in inflexible solutions that hinder growth and expose them to significant property risks.

Chapter 5 (Shiem-shin Then) continues along the same lines of evaluation, with a considered approach to space planning. Space expressed as square metres is the unit of measure: the challenge becomes one of balancing supply and demands — both today and in the future.

1.9.3 Adjust

Once the change requirements and the extent of the task has been identified, the essence of a 'project' starts to materialise. This is captured in Chapter 6 (Love and Smith), which considers the *strategic* or early design stages in the life of a project which determine the fundamental characteristics of quality, cost and time of projects. The inception stage provides an opportunity for *breakthrough*, innovative or creative ideas or just better approaches to facilities provision. The authors of this chapter advocate the creative engagement of a diverse, broad or mixed environment of stakeholders at this vital stage.

Pre-design evaluation is a concept addressed in Chapter 7 (Ornstein and Andrade). The name might suggest that it belongs in the 'evaluation' stage, but it is also instrumental in the fine-tuning of design solutions — often providing vital insights that need to be captured in the early design (inception) stage.

1.9.4 Carry out

The terms 'mobilisation' or 'implementation' aptly describe this 'doing' stage. The physical move or the mobilisation of new service providers are what traditionally we have associated with facilities management. The 'doing' stage is fraught with dangers associated with commissioning, testing and training. Things invariably go wrong, despite the best laid plans.

Chapter 8 (Bull) isolates 'communication' as the single most important determinant of success at this stage. Employees involved in a move are likely to be significantly more forgiving of mistakes if they understand the challenges involved and their role in the process (as co-producers). Timely communication provides the vital mechanism for allaying fears, tackling frustrations and resolving outstanding issues.

The theme of user-involvement is further explored in Chapter 9 (Vischer). Participative involvement of users is a phenomenon which looks set to transform the facilities management profession. The principles espoused in this chapter inform all of the stages of the REACTT model, from recognition to tracking and beyond. However, user involvement is of vital importance at the 'carry out' or implementation stage. The chapter highlights the possibility of transforming end users negative energy (associated with a resistance to change) to a powerful positive energy.

1.9.5 Track

Change management is a continuous process. Once the project is complete, many lessons remain to be learned. Often it is only by occupying the new space or experiencing the new service are shortcomings in the original design apparent. Added to this is the fact that even from day one of handover, requirements may have changed. Only by undertaking a 'post occupancy evaluation' some months after project completion can we identify shortcomings and embrace successes.

Chapter 10 (van der Voordt *et al.*) looks in detail at the concept of post occupancy evaluation, providing an overview of the principles involved. Only by continuing to track the results of a change management initiative can we provide the learning experience essential to the facilities manager.

1.9.6 Treasure

The REACTT model is a continuous cycle that occurs at many levels and across many time-frames in facilities management. The inevitability of 'obsolescence' is not assumed in this model. Indeed, buildings can perform well beyond their intended design life and often acquire a value that transcends any functional analysis. The concept of 'treasuring' what we possess runs through the final two chapters of the book.

In Chapter 11 (Inalhan and Finch) we examine how building users learn to value the familiar. Sometimes this is based on sound reasoning related to familiarity with processes, investment of effort in the development of expertise and asset specificity (having a unique knowledge of the peculiarities of a given system). However, we also need to consider the emotional process of 'place attachment'. Facilities managers ignore the strength of this phenomenon at their peril. The chapter examines the dimensions of place attachment that have to date been poorly understood in facilities management, despite the extensive research that abounds in other areas such as tourism, home-making and urban planning.

Chapter 12 (Pereira Roders and Hudson) is the final chapter of the book, lifting the lid on the concept of 'cultural heritage'. What is conveyed in the chapter is that in order to successfully manage change we also need to include assessment processes that recognise the existence of cultural heritage. Cultural heritage in this context includes the range of characteristics that are deemed to be of cultural significance. These can include tangible heritage such as buildings, engineering structures, archaeological sites and historic areas. They can also include intangible heritage assets such as customs, events and associations. In a world in which sustainability forms such a prevalent part, it is undoubtedly true that facilities managers need increasingly to understand how these assets are valued and treasured.

1.10 SUMMARY

This chapter has attempted to position facilities management in the larger context of organisational change. Today facilities management *is* change management. As such, the profession needs to be empowered with tools and techniques that enable them to engage with the new reality. Fundamental to this new reality is the empowerment of building users, engagement in complex service provider relationships, alignment with organisational strategy and empathy for assets that underpin the fabric of modern society.

REFERENCES

Barrett, P. and Baldry, D. (2003). *Facilities Management: Towards Best Practice*. Wiley-Blackwell.

Becker, F.D. (1990). *The Total Workplace: Facilities Management and the Elastic Organization*. CRC Press.

CEN/TC348 (2006). Facility Management – Part 1: Terms and Definitions, EN 15221-1.

Choi, T. (1995). Conceptualizing continuous improvement: Implications for organizational change. *Omega*, 23(6), 607–624.

Denison, D.R. and Mishra, A.K. (1995). Toward a theory of organizational culture and effectiveness. *Organization Science*, 6(2), 204–223.

Gersick, C.J.G. (1991). Revolutionary change theories: A multilevel exploration of the punctuated equilibrium paradigm. *The Academy of Management Review*, 16(1), 10–36.

International Council of Nurses (2010). Definition of Nursing. Available at: http://www.icn.ch/about-icn/icn-definition-of-nursing/ [accessed December 18, 2010]

MacCrimmon, K.R. and Wehrung, D.A. (1988). *Taking Risks: The Management of Uncertainty*. New York: The Free Press.

Mintzberg, H. and Westley, F. (1992). Cycles of organizational change. *Strategic Management Journal*, 13, 39–59.

Royal Institute of British Architects (1999). Outline Plan of Works, Services Supplement: Design and Management, in *Standard Form of Agreement for the Appointment of an Architect*. London: RIBA; 1999. SFA/99. Royal Institute of British Architects.

Tushman, M.L. and Romanelli, E. (1985). Organizational evolution: A metamorphosis model of convergence and reorientation. In *Research in Organizational Behavior*. pp. 171–222.

Williams, B. (1996). Cost-effective facilities management: a practical approach. *Facilities*, 14, 26–38.

2 Change Readiness

Edward Finch

CHAPTER OVERVIEW

In Shakespeare's character Hamlet we see a man overwhelmed by the political flux and uncertainties in his kingdom. He makes the profound observation that 'the readiness is all'. Much the same could be said about the readiness of facilities to support change. Some of the earliest work on change management by Coch and French (1948) dealt with the idea of 'creating readiness'. Their concern was with reducing employee's resistance to a change that is perceived as being imminent within an organisation. Readiness can be defined as a 'willingness or a state of being prepared for something' (Cambridge, 2008). The human dimension becomes more apparent in the implementation stage and is considered in more detail in Chapter 9 on the subject of user engagement. But readiness is not just a human property, it is an organisational property and a property of facilities themselves.

This chapter tackles the issue of 'readiness' in two respects: (1) the 'willingness' of a facilities management (FM) enterprise to respond to change and (2) the 'state of being' of physical assets to respond to change (flexibility). Both of these challenges are more than simply overcoming or avoiding resistance to change. The art of change readiness depends on the ability to foresee future (social, economic, technical) changes from the external environment. Such scenario planning requires an understanding of a number of possible futures — probable, likely and improbable. Only by understanding the multiplicity of future states is it possible to adopt an appropriate facilities solution and the supporting facilities management solution. Without such a vision, buildings become over-specified for improbable futures and under-specified for futures which could have been anticipated. Not only can the building itself entomb organisations; inflexible contracts involving service providers can result in a costly misfit with requirements that becomes more prominent with time. This chapter considers both of these issues in turn — the challenge of flexible relationships in service provision (soft-FM) and the challenge of flexible facilities (hard-FM).

Keywords: Change readiness; Flexibility; Adaptability; Responsiveness; Layering; FORT model.

Facilities Change Management. Edited by Edward Finch.
© 2012 Blackwell Publishing Ltd. Published 2012 by Blackwell Publishing Ltd.

2.1 SERVICE PROVIDERS AND PARTNERING

How is facilities management best delivered in an organisation subject to rapid change? Debate on the matter often centres on whether to outsource or not: that is to say, whether to use a contract to employ an outside service organisation or whether to employ your own facilities management team. In reality, the in-house versus outsourcing debate has today moved on. Contracts are increasingly complex and the question is not just 'whether to outsource?', but 'how to outsource?' More specifically, what is the 'best fit' between the client organisation (and its capabilities) and the outsourcing provider in an increasingly unpredictable future? With increased complexity comes increased interdependency: the fate of the client and the service provider becomes entwined. Business failure of the appointed service provider can no longer occur without major ramifications for the client. With this in mind, the outsourcing relationship increasingly becomes one of mutual support and nurturing. The relationship itself needs to accommodate change both within the life of an individual contract but also in the life of the relationship. Such a relationship may undergo significant change as the service provider acquires an increasing diversity of professional capabilities.

2.2 OUTSOURCING RELATIONSHIPS

One of the most influential studies on the nature of outsourcing relationships is the 'Four Outsourcing Relationship Types' proposed by Kishore *et al.* (2003). Their study is based on a framework that classifies client–provider relationships. This relationship invariably changes over time. By understanding the principal characteristics of an outsourcing relationship, the model helps in understanding this change.

Using the Four Outsourcing Relationship Types (FORT) approach we can identify two key dimensions, as shown in Figure 2.1:

1. substitution: the degree of ownership substitution that the service provider assimilates
2. strategic importance: the strategic impact of the outsourced services.

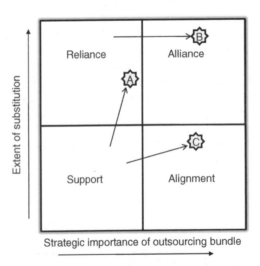

Figure 2.1 Four outsourcing relationship types (adapted from Kishore, 2003).

The first of these dimensions describes the amount of hardware (physical assets) and software (processes, intellectual know-how, expertise) that has been transferred to the provider. This includes 'asset specific' knowledge that is acquired through familiarity with the particular building portfolio. Such experiential knowledge cannot be easily re-acquired by the client or captured with the intention of transferring it to other competing providers at a later date. Added to this is the learning curve undergone by the service provider in relation to the functioning of the organisation, often developed over several years.

The second of the dimensions considers the extent to which the outsourced portfolio adds value to key business processes. This may include enhancement of customer relations, improved supplier relationships, improvement of product or service offering among others. This strategic sensitivity comes under careful scrutiny when considering the outsourcing step as it is this dimension that impacts on core business.

The FORT model of Kishore *et al.* (2003) uses a series of concise questions to identify (1) the level of substitution and (2) the strategic impact. An example question for establishing level of substitution is:

> To what extent are service providers involved in the planning, development, and implementation of new application systems for the client firm? (Kishore (2003, p. 88))

An example of a diagnostic question to assess the strategic importance of an outsourced portfolio is:

> To what extent does the outsourced portfolio call for a close partnership between the provider and the client firm? (Kishore (2003, p. 88))

Based on the scores emerging from the full set of distinguishing questions along both dimensions, the model categorises outsourced relationships into one of four types.

- Support: a relationship with low strategic impact and low levels of substitution. This type of relationship involves the lowest level of set-up costs and switching costs. Suited to less complex outsourcing tasks, the effective implementation of the support approach relies on ongoing monitoring (day-to-day) and reliance on benchmarking data.
- Alignment: in the alignment relationship, the level of substitution remains low. However, the strategic impact of the outsourced portfolio is high. This relationship typically represents a 'project based' contract involving a consultancy. In this scenario, the role of the 'intelligent client' becomes more prominent, providing technical expertise about the existing systems and the implementation of a new system.
- Reliance: in this scenario, the extent of substitution is high, whilst the strategic significance of the portfolio is relatively low. The common driver for this type of relationship is cost reduction. An *outcome*-based approach to control is more effective than behaviour-based control. However, this may be augmented by incentivisation schemes. The client organisation needs to be mindful of the risks of losing in-house expertise, particularly in relation to tendering and defining performance specifications.
- Alliance: evident in this type of relationship is both high levels of substitution and high strategic impact of the outsourced portfolio. This is the most comprehensive form of outsourcing (sometimes referred to as Total Facilities Management). In such a relationship there is a high reliance on mutual trust. Monitoring on a day-to-day

basis is entrusted to the service provider. Profit sharing is more in evidence, with the encouragement of mutually beneficial behaviours and shared intellectual property. Outcomes are measured in terms of behaviours rather than in terms of strict controls and outcomes.

How does change impact on outsourced service providers? The provider is often both an agent of change and may itself be the subject of change. Figure 2.1 illustrates possible evolutionary paths such organisations might take. This evolutionary path may well take place within the context of a single client–provider relationship. However, it is unlikely to be realised within the constraints of an existing contract which largely determines the type of relationship that is possible. Like other types of change, the desired cultural shift required may be met by forces of inertia, both within the service provider organisation and in the client organisation.

One example of change is the transition from *support* to *reliance*. This is shown in Figure 2.1 as a transition to point A. In such a scenario the client entrusts the service provider with greater responsibility and looks to them as a sources of capital injection to support upgrading measures (e.g. energy-efficient plant). As such, the client may operate more as a 'thin client' relying on leaner operation with less management responsibility. However, the continued reliance on an outcome-based model still requires a significant monitoring commitment. In return, the service provider is able to look at value-adding opportunities as well as performance-based rewards.

The transition from *reliance* to *alliance* represents a major change in the outsourcing relationship, as shown by destination B in Figure 2.1. Pivotal to such a transition is the move from an outcome-based mode of operation to one where performance is measured in terms of behaviours. Trust forms a central platform of this new approach and may be difficult to achieve in organisations that are accustomed to control and ongoing measurement.

An alternative path to B for a *support* service provider is path C, whereby the portfolio of services becomes more strategic in nature. In this context, the extent of substitution remains low but the value-adding possibilities offered through consultancy and project work is expanded. Familiarity with the client's specific assets and ways of working underpin this relationship. However, the ability to move in this direction depends on the ability of the provider to manage risks and develop competencies.

2.3   THE FM SUPPLY CHAIN

Figure 2.2 shows how the various service provider models (support, alignment, reliance and alliance) feed into the process of facilities change management. If your organisation is contemplating a journey of transformation, it is key to understand the type and capability of potential providers. Change management requires an understanding of the larger context, but all too often facilities managers are not entrusted with the full strategic agenda and are often left to rue the mistakes of others. Central to the success of any change management programme is the 'intelligent client'. As discussed in Chapter 1, this small team of individuals should be capable of securing (1) the organisational support (sponsorship), (2) the intelligence (reporting systems) and (3) the most effective delivery mechanism (service providers). The penultimate part of the jigsaw is the business unit. Informed by the overall strategy of the organisation, it is the projections and forecasts that emerge from each

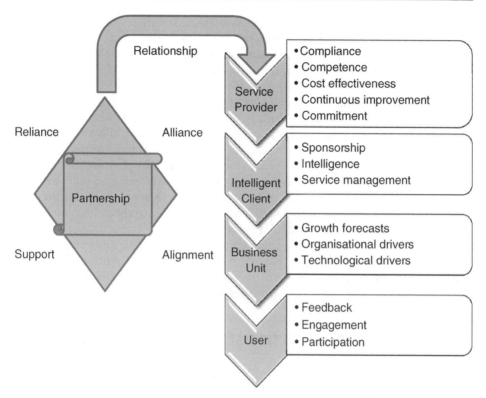

Figure 2.2 The facilities management supply chain.

of these that dictate the eventual shape of facilities provision, both in terms of size and nature. Furthermore, it is the desired culture and behaviours within each that need to be considered in a service agreement.

Figure 2.2 also highlights the increasing role of the *user* in facilities management. No longer is it sufficient to apply an autocratic, top-down, approach to the process. Users are co-producers and their cooperation will determine the level of change readiness. Only through consultation with users is it possible to tackle 'deep structures' that facilities management confronts. The outsourcing relationship strategy pivotally affects the type of user relationship that is achievable. A relationship based purely on competitive price (support model) is unlikely to have the flexibility to accommodate user involvement where it is required. Readiness is in part determined by the shared understanding and shared motivation of the service provider and user alike.

2.4   FLEXIBILITY IN SUPPORT OF CHANGE READINESS

Change readiness is not confined to individual, organisational or inter-organisational relationships. The infrastructure itself may possess the quality of change readiness. Buildings are capable of adjusting to the evolutionary changes of an organisation. A good building design can adapt in response to changing circumstances. This in-built capacity, which we describe as flexibility, refers to a facility's capacity to support change readiness. However, the dilemma for the designer is in anticipating likely changes. As such, the designer acquires the role of 'futurologist', 'technological forecaster' or 'foresight analyst'.

How can designers assist facilities managers in producing a more 'pliable' design solution with change readiness in mind? It is tempting to suggest that buildings should be designed for all eventualities. However, the concept of 'universal flexibility' is both technically and economically unachievable. Design for flexibility requires an under-standing of multiple future states, both possible and probable. Nowhere is this more apparent than in the arena of office design. Currently, flexibility is achieved through over-specification with respect to mechanical and electrical plant sizing, floor area provision and floor loading. The consequences of such over-provision are reduced efficiencies of plant, high occupancy costs and unnecessary maintenance. For some mission-critical facilities this may be seen as a price worth paying. In such situations the possible cost of disruption caused by intractable design characteristics (e.g. insufficient cooling capacity) soon justify the added investment in increased capacity. But in a world of environmental responsibility, this approach to flexibility presents a dilemma – simply 'building for growth' is both costly and unimaginative. In the era of sustainable design, facilities managers are looking for more ingenious approaches to flexibility: approaches that more reliably reflect the future demands of buildings.

2.5 BUILDING DESIGN DECISIONS AND FLEXIBILITY

Flexibility has become an overriding concern for organisations functioning in turbulent environments. Space provision has often failed to keep apace with the demand for rapid organisational change. Indeed, the ownership of real estate and the prevalence of long lease periods are often seen as impediments to organisational change (Harris, 2001). The *layered* model developed by Duffy (1990) provides an invaluable tool to guide decision makers in the 'flexibility conundrum' shown in Table 2.1. This layered approach to design decision making highlights the 'emergent' nature of buildings, made up of lifecycles within lifecycles. An analogy here is the human body: whilst the body itself may have an expected life beyond 70 years, organs such as the human skin may replenish themselves several times during that lifetime.

Table 2.1 Layered model of building systems. Duffy, F., (1990). Measuring building performance. Facilities, 8(5), pp. 17–20.

Building element	Decision-making lifecycle	Decision-making criteria
site	indefinite	location, orientation, neighbourhood, access
shell	50–75 years	shape, size and adaptability to organisational and technical changes
skin	25 years	aesthetics, integrity, energy efficiency
services	10–15 years	provision of cooling, heating, cabling and power capacity
scenery	5–7 years	describes the internal elements such as ceilings and partitions; these are tailored to organisational needs
systems	3 years	adapted to meet organisational processes and products and involves accommodation of ICT (information and communication technology)
settings	day to day	day to day arrangement of furniture and equipment

In the layered model the component layers include *site* (locality), *shell* (foundations, structure), *skin* (roof, external cladding), *services* (mechanical and electrical), *scenery* (internal partitions and ceilings), *systems* (IT) and *setting* (layout and furniture), as shown in Table 2.1. The separation of 'building structure' and 'fit-out' form a key part of this construct, with it-outs being revisited every time a new building tenant occupies the space at the beginning of a lease.

The significance of this model is that some design decisions are more intractable than others. For example, the choice of site remains fixed, once a location decision is made. Equally, the building shell cannot be easily modified during the lifetime of a building, which typically exceeds 50 years. Thus, for the client, any long-term commitment to the building must involve careful consideration of the constraining factors imposed by the building shell. In contrast, scenery, systems and settings are capable of being routinely altered, thus accommodating changing client requirements.

Not only does the model have implications for the appropriate decision-making cycles, the economic implications are also profound, as observed by Duffy:

> In a new building, costs are roughly divided into thirds - one-third for the 50-year shell, one-third for the 15-year services and one-third for the 5-year scenery. Add up what happens when capital is invested over a 50-year period: the shell expenditure is overwhelmed by the cumulative financial consequences of three generations of services and 10 generations of scenery. What appears to be so important in conventional building terms — the long term shell, foundations, walls, roof and structure — turns out to be nugatory in comparison to the gradual accretion of huge expenditure on ductwork and furniture. (Duffy (1990, page 18))

2.6   TYPES OF FLEXIBILITY

Flexibility means different things to different people — and therein lies the problem. Without a shared vision of future requirements, the concept becomes nebulous. The manufacturing sector has been more progressive in articulating different concepts of flexibility. A comprehensive classification system for the industry has been proposed by Sethi and Sethi, as shown in Table 2.2.

Table 2.2   A typology of manufacturing flexibility. Sethi, A.K. and Sethi, S.P., (1990). "Flexibility in Manufacturing: A survey", The International Journal of Flexible Manufacturing Systems 2, 289–328.

Flexibility category — manufacture	Description of flexibility
machine flexibility	the different operation types that an individual user/machine can perform
material handling flexibility	the ability to move the manufactured products within a manufacturing facility
operation flexibility	the ability to produce a product in different ways
process flexibility	the set of products that the work system can produce
product flexibility	the ability to add new products in the system
routing flexibility	the different routes (through machines and workshops) that can be used to produce a product in the system
volume flexibility	the ease to profitably increase or decrease the output of an existing system
expansion flexibility	the ability to build out the capacity of a building/system
program flexibility	the ability to run a system automatically
production flexibility	the number of products a system currently can produce
market flexibility	the ability of the system to adapt to market demands

Table 2.3 A typology of facility flexibility and change readiness (Finch, 2009). Reproduced by permission of the Associação Nacional de Tecnologia do Ambiente Construído.

Workplace (Finch, 2009)	Description of flexibility	Relevant layers, based on Duffy's categorisation (1990)
task flexibility	the diversity of 'activity settings' supported in the space	services, scenery, systems, setting
circulation flexibility	the ability to reconfigure circulation routes (vertical and horizontal circulation) to match organizational and individual requirements	shell, services, scenery
work style flexibility	the ability to undertake work in different ways	settings
process flexibility	the set of services that the work environment can support	services, scenery, settings
service flexibility	the ability to add new core services within the facility	services, systems, settings
space configuration flexibility (physical flexibility)	the different configurations that can satisfy the proximity requirements of the organisation	systems, settings
capacity flexibility	the ease with which to profitably increase or decrease the output of an existing system (e.g. number of hospital beds)	services, scenery, systems, setting
expansion flexibility	the ability to increase the capacity of a building/system	site, shell, services
building intelligence/ automation flexibility	the ability to run a facility automatically	services, systems
functional flexibility	the number of services an environment can currently support	services, scenery, systems, setting
property portfolio flexibility	the ability of the property portfolio to adapt to market demands (resilience)	site, shell, skin, services

A similar categorisation appropriate to facilities management has been proposed by Finch (2009). Each of the categories is associated with a descriptor, as shown in Table 2.3. The final column of the table attempts to identify where in the *layering* model the specific flexibility category arises.

This typology removes ambiguity when applying the concept of flexible solutions. Some of the categories relate to capacity, others to diversity and yet others to future proofing. Used in conjunction with organisational scenario planning each aspect of flexibility can be prioritised and evaluated for each design solution.

2.7 CONCLUSIONS

In the current climate of dramatic organisational change, inflexible buildings are undoubtedly producing dysfunctional organisations. The concept of change readiness attempts to pre-empt change. Rather than simply taking responsibility for implementing change, the facilities manager seeks to remove barriers for change phenomenon that are yet to be realised. This chapter has seen that 'willingness or a state of being prepared for something' can minimise the disruption that change brings. The inflexibility of traditional contracting can be superseded by more responsive relationships. Facilities can be engineered to accommodate possible futures. What is most apparent from our analysis of change readiness is that the facilities management role is increasingly one of futurology. More than that, it can no longer rely on historic forecasts. The future increasingly shows little relationship with the past. Have you ever tried to drive a car forward by looking in the mirror? Yet, most modern day techniques in facilities management, including budgeting and space forecasting, are based on historic projections. Change readiness demands that facilities be looked at with a fresh pair of eyes.

REFERENCES

Cambridge (2008). *Cambridge Advanced Learner's Dictionary.* Cambridge University Press.

Coch, L. and French, J. (1948). Overcoming resistance to change. *Human Relations*, 1, 512–532.

Duffy, F. (1990). Measuring building performance. *Facilities*, 8(5), 17–20.

Finch, E. (2009). Flexibility as a design aspiration: the facilities management perspective. *Ambiente Construído*, Vol. 9, No 2.

Harris, R. (2001). From fiefdom to service: The evolution of flexible occupation. *Journal of Corporate Real Estate*, 3, 7–16.

Kishore, R. *et al.* (2003). *A relationship perspective on IT outsourcing. Communications of the ACM*, 46(12), 86–92.

Sethi, A.K. and Sethi, S.P. (1990). Flexibility in manufacturing: A survey. *The International Journal of Flexible Manufacturing Systems*, 2, 289–328.

3 Form, Function and the Economics of Change

James Pinder, Simon Austin, Rob Schmidt III, and Alistair Gibb

CHAPTER OVERVIEW

Chapter 1 explored the forces and pace of change facing organisations and the implications for those responsible for managing their buildings. This chapter looks in more detail at the relationship between buildings and change, and examines how this relationship can be managed. In doing so, it provides a foundation for Chapter 4, which looks at how we can prepare for possible scenarios based on change readiness.

This chapter begins by looking at a way in which buildings can be a catalyst or constraint to change, both physically and symbolically. It then goes on to look at the impact of changing demands on building performance and how this is manifested in terms of obsolescence — the operational costs and constraints borne by occupiers — and depreciation — the reduction in rental income and capital values experienced by building owners.

The third and fourth parts of this chapter discuss two interrelated approaches for coping with changing user demands: designing buildings for adaptability and adaptive re-use. This chapter concludes by discussing the implications of obsolescence, depreciation and adaptability for facilities managers and the importance of maintaining a feedback loop between facilities management and design.

Keywords: Adaptability; Adaptive re-use; Building performance; Depreciation; Obsolescence.

3.1 INTRODUCTION

In recent years there has been growing recognition of the impact that buildings can have on organisational change. In a world characterised by rapid social, technological, economic and political change, buildings can serve to constrain organisations by acting as a physical or symbolic anchor to the past. However, the physical and symbolic nature of buildings also means that they can be used to support and facilitate change. For example, Mawson (2006) highlighted the case of Marks and Spencer (a major UK high street retailer), which used its move to a new headquarters building in 2004 to encourage a more flexible and open

Facilities Change Management. Edited by Edward Finch.
© 2012 Blackwell Publishing Ltd. Published 2012 by Blackwell Publishing Ltd.

organisational culture, characterised by increased creativity and rapid decision making. Similarly, in the UK's public sector a number of government bodies have used changes to their physical working environment as an opportunity to introduce new working practices and business processes (Allen *et al.*, 2004; Hardy *et al.*, 2008).

A key part of understanding the impact that buildings can have on change is an understanding of the impact that change can have on buildings and the ways in which buildings can accommodate change. As well as being an important operational resource for many organisations, buildings represent a significant investment in capital and physical resources — so ensuring that they can adapt to change is imperative, from both a financial and an environmental perspective. In this chapter we will explore the impact that changing demands can have on the performance of buildings and the implications for owners, occupiers and facilities managers. We will also discuss the different design strategies that can be employed to enable buildings to adapt to changing demands, together with illustrative examples from a range of sectors. In the final part of this chapter we will look at the opportunities for adapting existing buildings to accommodate changing demands and how facilities managers can help to overcome the barriers to adaptive re-use.

3.2   CHANGING DEMANDS

When buildings are carefully designed, constructed and maintained, their physical life spans can be almost indefinite (Ashworth, 1997). A building's performance will decline as the condition of its structure and fabric deteriorates, but its physical life can normally be extended through periodic maintenance and refurbishment activity. However, even when a building remains in good physical condition, rising expectations can serve to reduce its service life (Figure 3.1). The *service life* refers to the period of time over which a building functions above a minimum acceptable level of performance (Iselin and Lemer, 1993). This reduction in a building's service life due to rising expectations is called 'obsolescence', the causes of which can be wide-ranging, including changes in legislation, technology, economic conditions or architectural style (Mansfield and Pinder, 2008).

One of the most succinct explanations of obsolescence was by Burton (1933), who suggested that if it were possible to hold stationary the physical condition of a building, obsolescence would be the difference between the existing demand for the characteristics of the building and the demand that was anticipated when the building was originally constructed. The building in question may not necessarily be dilapidated or worn out, although these factors may accentuate the obsolescence: the building simply does not measure up to contemporary expectations (Iselin and Lemer, 1993). Building obsolescence therefore constitutes a relative, rather than absolute, decline in performance — two buildings of the same type in the same location may exhibit different levels of performance because of differences in their physical characteristics: one building will be less obsolete than the other (Nutt *et al.*, 1976).

Three interrelated factors mean that buildings are particularly vulnerable to the impact of obsolescence.

First, and foremost, most buildings have a fixed location, which means that they cannot be moved to accommodate changing demands (Raftery, 1991). Location is usually the key driver behind the utility and value of a building, such that any change in either the characteristics of an area or its attractiveness to occupiers, relative to other locations, will impact upon the level of demand for that building. The impact of locational factors on

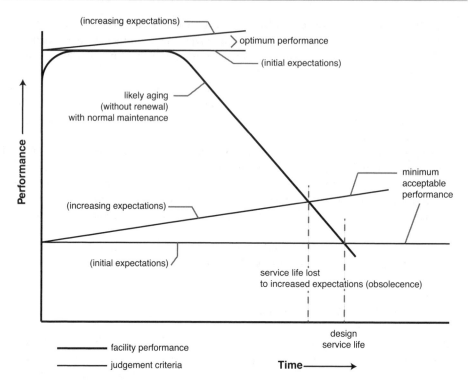

Figure 3.1 Conceptual view of building obsolescence. Reproduced by permission from Iselin and Lemer, 1993 by the National Academy of Sciences, Courtesy of the National Academies Press, Washington, D.C.

obsolescence are usually 'incurable', in the sense that it is not normally feasible for individual building owners to improve the infrastructure of an area or reverse urban decay. Normally changes on this scale can only be realised with co-ordinated investment and long-term planning by local authorities and other government agencies (Gann and Barlow, 1996; Heath, 2001).

The second factor that gives rise to obsolescence is that most buildings are designed to be durable, which means that they tend to outlive the design standards, building regulations and user requirements that prevailed at the time of their construction (Ohemeng and Mole, 1996). For instance, a 2004 study of the UK's commercial and industrial building stock found that 53% of buildings in the UK were constructed before 1940 and a further 21% before 1970 (OPDM, 2004). However, this durability can also mean that a building that has become obsolete because of changing demands may benefit from further change (Weatherhead, 1999). For example, in the UK some office buildings that became redundant during the 1980s because they were unable to accommodate new information and communications technologies have since come back into use because of further techno-logical developments, such as wireless networks.

A third factor that contributes to the risk of obsolescence in buildings is that most are designed for a specific need or purpose. The capital cost of developing new buildings means that their 'opening configurations' (Iselin and Lemer, 1993) tend to closely reflect the known needs of their first group of users, rather than the often unknown needs of potential future users (Nutt, 1988), due to concerns about compromising the known needs of the former by attempting to accommodate the presumed needs of the latter. In larger property developments, this short-term thinking can be exacerbated by the time lag in the construction and development process, which means that it can take years between the initial planning and design of a new building and its completion and occupation, during which time user expectations and market demand may have changed significantly. This lag between demand and supply is also a contributory factor to the oversupply of buildings during cyclical downturns in the property market (Ball, 1994; Grenadier, 1995).

In the UK, interest in the issue of building obsolescence has tended to increase during periods of rapid economic and technological change. For instance, during the 1960s a number of studies looked at the issue in the context of the widespread urban regeneration that was taking place in many towns and cities at the time. The issue came to the fore again in the 1980s, when a growing number of 1950s and 1960s office buildings became obsolete due to changes in technology and business practices, particularly in the newly deregulated financial services sector. There was a growing feeling at the time that obsolescence had not been factored into property investment decisions (Baum, 1991), so that, as long-term leases ended, some commercial property owners were left with office buildings that had depreciated in value and required extensive refurbishment in order to meet new occupier requirements (Bryson, 1997). For some owner-occupiers, the decline in property values meant that office buildings that were once valuable assets became financial liabilities (Gibson, 1994).

The financial impact of the 1980s property slump in the UK prompted a number of studies into the impact of building obsolescence on depreciation, the most notable of which was undertaken by Baum (1991 and 1994). As part of his research, Baum asked occupiers in the City of London which office building characteristics were most important in determining rental value. He found that configuration, internal specification and external appearance were rated as the most important factors, although there were differences in the importance attributed to particular sub-factors (Table 3.1). For instance, with regards to building configuration, floor layout was seen as more important than floor-to-ceiling height. Baum suggested that uncertainty over future occupier requirements meant that investors should look to purchase buildings that were flexible with respect to these characteristics, concluding that 'flexibility reduces the risk of an irreversible and major reduction in the market value of a building' (Baum, 1994; p. 39).

The idea of designing buildings to be flexible was by no means new: in the 1960s Weeks (1963 and 1965) talked about 'indeterminate architecture' and in the 1970s Alex Gordon, the then President of the Royal Institute of British Architects, put forward the principle of 'long life, loose fit and low energy' buildings (Anon, 1972; p. 26). However, in the 1980s and 1990s many UK property developers reacted to the demand from institutional investors for greater flexibility by over-specifying, amongst other things, comfort cooling services, small

Table 3.1 Importance of office building sub-factors. Baum, A. (1994). Quality and property performance, Journal of Property Valuation & Investment, 12(1), pp. 31–46.

Factor	Rating	Sub-factor	Rating (percentage)
Configuration	1	Floor layout	86.4
		Floor-to-ceiling height	13.6
Internal specification	2	Quantity/quality of services	79.7
		Quality of finishes	20.3
External appearance	3	Impact of entrance hall, etc ...	72.1
		Quality of external design	27.9
Deterioration	4	Deterioration of interior	65.1
		Deterioration of exterior	34.9

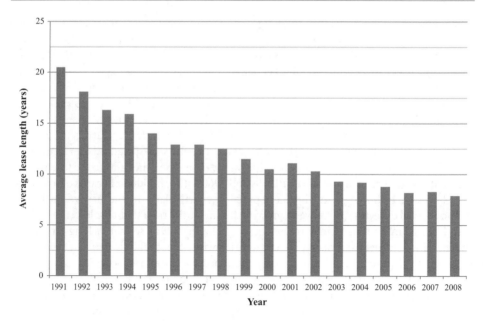

Figure 3.2   Average lease lengths in the UK, for all property types (compiled from BPF and IPD, 2001, p. 7. and 2009, p. 4).

power provision and floor loadings, in order to try to 'future proof' their office buildings and cater for potential changes in occupier demand (Stanhope, 1992 and 2001), even though these levels of redundancy were rarely required. This resulted in more expensive and more energy-intensive office buildings, both in terms of the energy embodied in the buildings' fabric and the energy used in operating the buildings (Guy, 1998).

Over the last two decades falling lease terms in the UK have underlined the need for greater flexibility in the commercial building stock (Ellison and Sayce, 2007). Traditionally, the UK property market was characterised by the use of 25-year lease terms with upwards only rent reviews, which provided property owners and institutional investors with greater financial security. However, whilst these 'institutional leases' were good from an owner's or investor's perspective, they were usually too inflexible for occupiers, particularly those with rapidly changing property needs (Lizieri, 2003). Since 1990, when the average lease term for office buildings was 23.5 years in the UK (Hamilton *et al.*, 2006), there has been a steady fall in average lease lengths across all sectors (Figure 3.2), with the average length of new leases standing at 7.9 years in 2008 (BPF and IPD, 2009). The likely consequence of this change is that buildings will be required to meet the needs of a variety of tenants over their life times, due to higher tenant turnover (Ellison and Sayce, 2007).

The demand for greater flexibility in building design has also been reflected in the widespread adoption of open-plan office environments in the UK and elsewhere. Although open-plan office environments had been around for decades in one form or another, since information technology became ubiquitous in offices in the 1980s and 1990s an increasing number of occupiers have moved away from traditional cellular office designs, which are seen as costly to change and a barrier to new ways of working (Longcore and Rees, 1996; Worthington, 2001). However, as with any flexible building design strategy, the use of open-plan office environments can necessitate trade-offs: in some cases, the same qualities that enable an office environment to be more flexible may give rise to higher levels of background noise and less privacy for the people working within it. If not managed effectively such trade-offs can have negative consequences for the users of open-plan office environments (de Croon *et al.*, 2005).

3.3   DESIGNING FOR ADAPTABILITY

Designing a building to be flexible is just one way of mitigating the risk of obsolescence and ensuring that it can adapt to changing needs. Research undertaken as part of the *Adaptable Futures* project at Loughborough University has identified six different strategies for designing adaptability into buildings (Table 3.2). The adaptable design strategies vary depending on the type of change that needs to be accommodated, ranging from a change of task on the one hand through to a change of location on the other. The decision level also varies between stakeholders from strategy to strategy. For instance, decisions about changes of task are more likely to be made by users, whereas decisions concerning a change in building scale or location tend to be taken by the building owner. Similarly, each adaptable design strategy is dependent on the capacity of different building layers[1] to accommodate change and involves different frequencies or rates of change.

Adjustable design strategies focus on ensuring that the 'stuff' inside a building, such as furniture, equipment and appliances, can easily be reconfigured to accommodate changing tasks. The frequency of change therefore tends to be high. Examples of adjustable design can be found in a number of sectors. For instance, in the pharmaceutical industry GlaxoSmithK-line (GSK) has developed its FlexiLab concept in order to help it to respond to rapidly changing research and development needs. Whereas previously GSK's scientists were impeded by fixed furniture, fume cupboards and services, FlexiLab enables the layout of their laboratories to be reconfigured quickly and at minimal cost through the provision of 'plug and play' services, moveable fume cupboards, storage and benching, and re-locatable glass partitions between the laboratory spaces and write-up areas. A key facet of adjustable design strategies such as FlexiLab is that users of a facility are provided with greater agency to influence the configuration of their environment.

One of the most common adaptable design strategies is versatility (or flexibility) to change the spatial layout of a facility. The Centre Pompidou in Paris is a good example of a versatile facility, with internal spaces that were designed to be extremely flexible and support a variety of uses. This was achieved by moving the mechanical services, circulation routes

Table 3.2   Adaptable building design strategies (adapted from Schmidt III *et al.*, 2010; p. 7).

Strategy	Type of change	Decision level	Building layer(s)	Frequency of change
Adjustable	Change of task	User	Stuff	High
Versatile	Change of space	User	Stuff, space	High
Refittable	Change of performance	Owner	Space, services, skin	Moderate
Convertible	Change of function	User/owner	Space, services, skin	Moderate
Scalable	Change of size	Owner	Space, services, skin, structure	Moderate/low
Moveable	Change of location	Owner	Structure, site	Low

[1] See Chapter 2 for a more in-depth discussion of building layers.

and structure to the exterior of the building (Silver, 1994). Another example is the Henry Ford College in Loughborough, which was designed to accommodate a wide range of training and exhibition activities in a large single-span building, with internal walls and partitions that are suspended from the superstructure. In most cases the versatility of a facility is dependent upon the adjustability of the 'stuff' contained within it. For instance, many contemporary 'open-plan' offices are intended to be versatile, but their adaptability is often constrained by the inability for users to easily adjust the location and configuration of the furniture and equipment within the space.

Refittable design strategies involve changing the performance of a building by altering its space, services or skin. An example of a refittable design strategy would be 'designing in' the potential to change a building's skin and services so that they can achieve improved levels of thermal performance in order to mitigate against the risk of obsolescence due to potential climate change, higher energy prices or new legislation. The ability of building owners to easily refit and change the energy performance of their buildings is likely to become increasingly important in the future, particularly as energy-efficient buildings become more attractive to occupiers and start to command a premium over 'hard to treat' (Jones, 2009; p. 381) buildings. For instance, in the UK the introduction of Display Energy Certificates means that it is easier for occupiers to differentiate between the energy performance of different buildings and factor this into their decision making (Guertler *et al.*, 2005).

For many people, adaptability is synonymous with changing the function or use of a building. Many buildings are converted to accommodate different functions, even though such a change of use was never envisaged when the building was originally designed and constructed. Examples in the UK include the Victorian mill buildings (Williams, 1985) and 1960s office buildings (Gann and Barlow, 1996) that have been converted to residential use. However, some buildings are specifically designed with change of function in mind. For instance, 3D Reid's Multispace concept was developed to enable building owners to alter the mix of space use in a development scheme to reflect different market conditions, but without having to alter the building's shell and structure (Davison *et al.*, 2006). Another example of designing for convertibility is SEGRO's Energy Park in Vimercate (Figure 3.3), near Milan, which was designed to be fitted-out as an office, warehouse, laboratory or storage space. As an investor-developer with a long-term interest in the buildings that it

Figure 3.3 Designed for change of function – SEGRO's Energy Park in Vimercate, near Milan.

develops, SEGRO sees adaptable design as a key part of its business strategy for responding to changing occupier demand for buildings.

A fifth adaptable design strategy involves changing the size of buildings. The notion of designing buildings to be scalable is particularly appealing to organisations with financial constraints, plans for growth or contraction, or uncertain future space requirements. For example, in the 1970s the National Health Service (NHS) introduced the Nucleus Hospital programme during a period of limited capital spending, the idea being that small-scale hospitals could be built (using a standard template of $1000 \, \text{m}^2$ blocks) to house a nucleus of departments which could then be expanded at a later date, as more capital funding became available. A total of 130 Nucleus Hospitals were built (Francis *et al.*, 1999). A more recent example of scalable building design is the Igus factory in Cologne, which was designed to be built in phases as the company's business grew. The factory, which is also made up of a series of standard blocks, has been expanded seven times over a 17-year period, from an original floor area of $4500 \, \text{m}^2$ to a current floor area of $20\,000 \, \text{m}^2$ (Fuster *et al.*, 2009). By contrast, the stadium constructed for the Sydney Olympics in 2000 was designed to be scaled down from 110 000 seats to 80 000 seats after the event (Searle, 2002).

In the UK at least, very few buildings are designed to be moveable. One of the most notable examples of a moveable building is the British Antarctic Survey's Halley VI Research Station, which was designed with legs and skis so that it can be periodically moved inland as the ice sheet on which it sits slowly moves out to sea (Abley and Schwinge, 2006). However, while truly moveable buildings are few and far between, it is not unusual for specific parts of buildings to be designed to be moveable. For instance, the Igus factory referred to above includes stand-alone office pods that can be moved around the factory's floor as business processes change (Fuster *et al.*, 2009). The Stade de France in Paris is another example of where a moveable design strategy has been adopted: although primarily used to host football and rugby matches, the stadium's first tier of seating can be retracted to expose an athletics track (Beech and Chadwick, 2004).

The degree to which the adaptable design strategies described above are used in practise depends on a wide range of factors, the interplay between which will be different from building to building. These factors include prevailing market conditions, planning restrictions, the design philosophy of the architects, procurement methods and the business model of the client or developer. For instance, a study of the Norwegian office market by Arge (2005) found that owner-occupiers and developers that were developing buildings to manage and let themselves tended to incorporate more adaptable design features into their buildings than developers that were developing buildings for sale. In a different context, Barlow and Köberle-Gaiser (2008) suggested that the use of the Private Finance Initiative (PFI) as a method of procuring healthcare facilities in the NHS had impeded the use of adaptable design solutions in the schemes that they examined, due to the complex structure of projects and the way that risks were allocated between the client and developer.

In recent years, interest in designing for adaptability has been spurred on by concerns about the environmental impact of buildings, the notion being that buildings that are more adaptable are more sustainable (Kendall, 1999; Graham, 2006). This was highlighted in a recent study undertaken in the UK by Ellison and Sayce (2007). They undertook focus groups with property investment and valuation professionals in order to find out which sustainability criteria were being factored into property investment appraisals and which were being included in calculations of worth. Overall, building adaptability was rated as the second most important sustainability criterion after accessibility, ahead of other sustainability criteria such as energy efficiency and water consumption (Table 3.3). The authors

Table 3.3   Sustainability criteria in rank order. Ellison, L. and Sayce, S. (2007). Assessing sustainability in the existing commercial property stock: Establishing sustainability criteria relevant for the commercial property investment sector, *Property Management*, 25(3), p. 290.

Criteria	Average score (/25)
Accessibility	15.04
Building adaptability	13.51
Pollutants	10.77
Contextual fit	10.32
Energy efficiency (including climate control)	8.39
Occupier	7.07
Waste management	5.42
Water consumption	2.13

attributed this finding to the fact that greater adaptability would reduce the need for building owners to undertake 'costly and environmentally damaging redevelopment' (Ellison and Sayce, 2007; p. 297).

The link between sustainability and adaptability of buildings has been underpinned by a growing awareness of the energy embodied in buildings, both during initial construction and during refurbishment and remodelling. For example, Yohanis and Norton (2002) estimated that the energy embodied in an office building during construction was equivalent to around 67% of its operating energy over a 25-year period (Figure 3.4), a

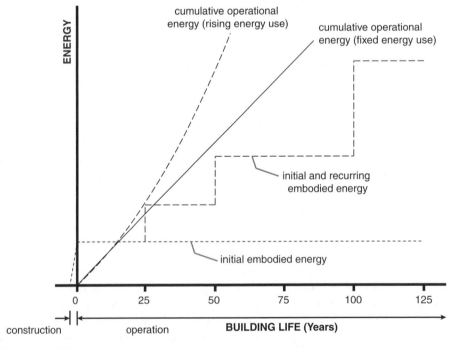

Figure 3.4   Estimated operational and embodied energy as a function of building life. Reprinted from Yohanis, Y. and Norton, B. (2002). Life-cycle operational and embodied energy for a generic single-storey office building in the UK, Energy, 27, p.78., with permission from Elsevier.

figure that was expected to increase in the future as energy use declines due to improved energy-efficiency measures. When additional (or recurring) embodied energy from periodic refurbishment and adaptation activity was included in their calculations, the cumulative embodied energy was estimated to be greater than the energy used in operation over the same 25-year period. Designing buildings to be more adaptable could therefore be beneficial by reducing the amount of embodied energy arising from the construction of new buildings and the adaptive re-use of existing buildings.

3.4   ADAPTIVE RE-USE

The term *adaptive re-use* is often used to refer to a change in the function of a building (Kincaid, 2000; Heath, 2001; Shipley *et al.*, 2006; Bullen, 2007 and Langston *et al.*, 2008). Indeed, much of the existing research into adaptive re-use has stemmed from a desire to find new uses for vacant office and industrial buildings, which in most cases has involved exploring the potential to convert such buildings to residential use (Williams, 1985; Barlow and Gann, 1995; Gann and Barlow, 1996; Coupland and Marsh, 1998; Freer *et al.*, 1999; Heath, 2001; Geraedts and van der Voordt, 2002; Remøy and van der Voordt, 2007). However, we use the term here to encompass the broad range of changes described above in Table 3.2. Hence, under this broader conceptualisation, refitting a building's skin and services to improve its thermal performance constitutes adaptive re-use, even if the building continues to accommodate the same function as before. Likewise, a building may be extended or reduced in size by changing its space, services, skin and structure, but the function of the building may nevertheless remain the same.

Although most buildings have not been designed with adaptability in mind, many have subsequently been adapted in one way or another to meet changing needs and expectations. Understanding how such buildings have been adapted and re-used – and why other buildings have remained vacant or have been demolished – can provide a useful insight into how adaptability can be designed into new buildings. For example, a recent study of 1960s university buildings in the UK by AUDE (2008) suggested that adaptive re-use was more likely to be successful in buildings with generous floor to soffit heights, good vertical access for services, frame structures with suitable grids and good vertical movement for building occupants. However, the potential for adaptive re-use may also be determined by factors external to the building itself; Gann and Barlow's (1996) research into the feasibility of converting empty offices into flats found that the capacity of a building to change depends on its surrounding infrastructure and amenities.

In the US, Slaughter (2001) analysed data from 48 building renovation projects with the aim of identifying design characteristics that facilitate adaptive re-use. Her analysis revealed that the inter-relationships between different elements of a building had a strong impact on its capacity to change, in that greater physical separation enabled one element to be changed without affecting another element. The importance of physical separation between different elements of a building can be illustrated through a case study undertaken as part of the *Adaptable Futures* project at Loughborough University. When the David Wilson Library (Figure 3.5) at the University of Leicester in the UK was constructed in the 1960s the building's services were integrated within its reinforced concrete frame, which at the time was a ground-breaking configuration that won the building a RIBA excellence award. However, when the building was later extended this configuration made changing the

Figure 3.5 Adapted for changing needs – the David Wilson Library at the University of Leicester.

building's services more difficult than it could otherwise have been, because the two building layers were in effect frozen together.

Property developers have traditionally considered adaptive re-use projects to be more expensive and more likely to suffer cost over-runs and programme delays than new-build projects (Egbu, 1995). The risks associated with adaptive re-use have been attributed to a range of factors, including:

- site or access constraints
- concerns about the potential discovery of asbestos or other harmful contaminants during building works or
- the time and effort required to obtain listed building consent or planning consent for a change of use.

Moreover, in many countries existing buildings that are adapted and re-used are required to comply with current building regulations, which for some types of buildings can prove costly and technically challenging, particularly when it relates to matters such as energy efficiency and disabled access. However, in recent years the negative perceptions of adaptive re-use have begun to be challenged.

For instance, Scott Brownrigg *et al.* (2009) found that even the comprehensive refurbishment of an existing building, involving the replacement of its façade and services, can be completed in a shorter time frame than a comparable new-build project (Figure 3.6). They attributed this difference to, amongst other things, the programme risks associated with groundworks and the construction of the sub-structure in new-build projects. The authors also found that adaptive re-use schemes were more cost effective than equivalent new-build projects, with comprehensive refurbishment schemes being on average 14% cheaper. Shipley *et al.* (2006) also explored the comparative costs of adaptive re-use and new-build projects, by comparing refurbishment cost data for 23 adaptive re-use 'heritage' projects with a typical new-build cost for buildings of a similar type and size. Their analysis

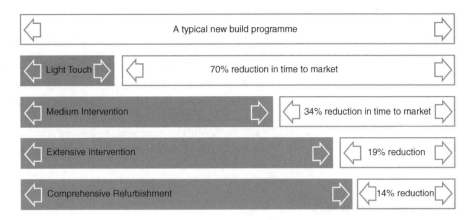

Figure 3.6   The programming benefits of refurbishment. Reproduced by permission of Scott Brownrigg Ltd.

suggested that adaptive re-use was more cost effective than new-build in medium-sized residential schemes and large-scale commercial projects. Moreover, the authors suggested that even in cases where adaptive re-use was more expensive than new-build, such as in large-scale residential projects, this was often compensated for by better financial returns, in the form of higher sale prices or rents.

One of the most in-depth studies of adaptive re-use was undertaken by Ball (1999), who examined the adaptive re-use of vacant industrial buildings in Stoke-on-Trent in the UK. The issue of vacant industrial buildings was a particularly significant issue in Stoke-on-Trent at the time of the study because of the de-industrialisation that had occurred in the area during the 1980s and early 1990s. As part of his research, Ball investigated the views of property developers regarding the constraints on adapting and re-using vacant industrial buildings and the benefits of adaptive re-use over redevelopment. He found that the perceived benefits of adaptive re-use derived from the good location of existing buildings, their solid build quality and character, and the lower cost of refurbishment compared with new construction. However, he also identified a number of constraints to adaptive re-use, including difficulties in gaining access to redundant buildings (due to absentee owners), planning constraints and inflexible building design.

Interest in adaptive re-use has grown in recent years as an increasing number of property owners and developers have begun to recognise the financial benefits of extending the life of existing buildings. This is particularly the case in the current economic climate, where capital spending is more limited and many organisations — particularly those in the public sector — are looking to make do with the buildings that they already have, rather than construct new ones. In addition, there has also been growing recognition of the potential non-financial benefits of adaptive re-use. These benefits vary from building to building, but may include the 'heritage value' of re-using existing buildings (Bullen, 2007; Wilkinson *et al.*, 2009), the fact the existing buildings are often situated in good locations (Shipley *et al.*, 2006) and the tendency for some older forms of construction, particularly those with greater a thermal mass, to facilitate more passive forms of heating and cooling (Langston *et al.*, 2008).

The growing acceptance of adaptive re-use as a viable alternative to new-build also reflects a change in mindset about the extent to which existing buildings can be changed and the range of functions that they can be used for (Ball, 1999; Shipley *et al.*, 2006). For example, when Stockport Primary Care Trust in the UK proposed adapting and re-using parts of Kingsgate House, a mixed-tenant 1960s town centre office building, to accommodate clinical services, both the building's landlord and the local planning authority were initially sceptical because this was seen to run counter to conventional wisdom about that type of uses that could be accommodated in an 'office' building. However, the resulting clinical environments are extremely popular with both staff and patients, because they are conveniently located and do not feel like traditional hospital spaces. The Trust has since adapted and re-used a number of other buildings in the town, including a former retail unit to house youth services and a former light-industrial manufacturing facility to accommodate drug and alcohol services.

3.5   CONCLUSIONS

This chapter has explored the inter-relationship between buildings and change, focusing on the impact that change can have on the performance of buildings and the ways in which buildings can be designed to accommodate changing demands. The capacity for buildings to accommodate change is determined primarily through design decisions taken early on in the their procurement, resulting in a building's design structure – what it is, how it is constituted (Baldwin and Clark, 2000). Designing for greater adaptability demands a shift away from the current emphasis on form and function, in response to immediate priorities, towards a time-based view of design. Architects tend to focus on aesthetics and functional performance, freezing out time in pursuit of a static idealised object of perfection (Brand, 1994; Hollis, 2009; Till, 2009). A reaction to this way of operating is the encouragement of a more dynamic and long-term view of the built environment (Habrakan, 1998), in which buildings are not finished works removed from time, but imperfect objects whose forms are in constant flux, continuously evolving to fit social, technological, and aesthetic changes in society.

A key prerequisite to designing buildings to accommodate change more readily – what might be construed as 'good design' – is developing a better insight into how existing buildings have changed over time and using this knowledge to the inform the briefing and design of new buildings. As the professionals responsible for managing and operating buildings, facilities managers can potentially play a key role in helping clients and their design teams to understand how to deliver buildings that cope better with changing demands. However, while the argument that facilities managers should input into the design briefing process is a longstanding one (Nutt, 1993; Brown *et al.*, 2001), in many construction projects this is still not the norm (Kelly *et al.*, 2005). As Kohler and Hassler (2002) point out, the existing building stock can be a valuable source of knowledge and learning for architects and engineers; the challenge for facilities managers is to create an effective feedback loop so that this knowledge and learning can be applied to the design of new buildings.

3.6   ACKNOWLEDGEMENTS

The authors would like to acknowledge the financial support of the Engineering and Physical Sciences Research Council and the Innovative Manufacturing and Construction Research Centre at Loughborough University, together with the input and case studies provided by the *Adaptable Futures* project partners. Further information about *Adaptable Futures* can be found at www.adaptablefutures.com.

REFERENCES

Ably, I. and Schwinge, J. (2006). Architecture with legs. *Architectural Design*, 76(1), 38−41.

Allen, T., Bell, A., Graham, R., Hardy, B. and Swaffer, F. (2004). *Working Without Walls: An Insight into the Transforming Government Workplace*. Office of Government Commerce, London.

Anon (1972). RIBA to probe use of buildings. *Design*, 283, July.

Arge, K. (2005). Adaptable office buildings: theory and practice. *Facilities*, 23(3/4) 119−127.

Ashworth, A. (1997). *Obsolescence in Buildings: Data for Life Cycle Costing*. CIOB Report No. 74, Chartered Institute of Building, London.

AUDE (2008). *The Legacy of 1960's University Buildings*. Association of University Directors of Estates, Cambridge.

Baldwin, C. and Clark, K. (2000). *Design Rules: The Power of Modularity*. MIT Press, Cambridge, Massachusetts.

Ball, M. (1994). The 1980s property boom. *Environment and Planning A*, 26(5), 671−695.

Ball, R. (1999). Developers, regeneration and sustainability issues in the reuse of vacant industrial buildings. *Building Research & Information*, 27(3), 140−148.

Barlow, J. and Gann, D. (1995). Flexible planning and flexible buildings: reusing redundant office space. *Journal of Urban Affairs*, 17(3), 263−276.

Barlow, J. and Köberle-Gaiser, M. (2008). The private finance initiative, project form and design innovation: The UK's hospitals programme. *Research Policy*, 37, 1392−1402.

Baum, A. (1991). *Property Investment Depreciation and Obsolescence*. Routledge, London.

Baum, A. (1994). Quality and property performance. *Journal of Property Valuation & Investment*, 12(1), 31−46.

Beech, J. and Chadwick, S. (2004). *The Business of Sports Management*. Pearson Education, Harlow.

BPF and IPD (2001). *Annual Lease Review*. British Property Federation and Investment Property Databank, London.

BPF *and* IPD (2009). *Annual Lease Review*. British Property Federation and Investment Property Databank, London.

Brand, S. (1994). *How Buildings Learn: What Happens After They're Built*. Penguin, New York.

Brown, A., Hinks, J. and Sneddon, J. (2001). The facilities management role in new building procurement. *Facilities*, 19(3/4), 119−130.

Bryson, J.R. (1997). Obsolescence and the process of creative reconstruction. *Urban Studies*, 34(9), 1439−1458.

Bullen, P. (2007). Adaptive reuse and sustainability of commercial buildings. *Facilities*, 25(1/2), 20−31.

Burton, J.E. (1933). Building obsolescence and the assessor. *The Journal of Land and Public Utility Economics*, 9(2), 109−120.

Coupland, A. and Marsh, C. (1998). The conversion of redundant office space to residential use. *Paper presented at the RICS Cutting Edge Conference*, September 2008, London.

Davison, N., Gibb, A., Austin, S., Goodier, C. and Warner, P. (2006). The Multispace adaptable building concept and its extension into mass customisation. *Paper presented at the International Conference on Adaptable Building Structures*, 3-5 July, Eindhoven, Netherlands.

de Croon, E., Sluiter, J., Kuijer, P. and Frings-Dresden, M. (2005). The effect of office concepts on worker health and performance: A systematic review of the literature. *Ergonomics*, 48(2), 119−134.

Egbu, C. (1995). Perceived degree of difficulty of management tasks in construction refurbishment work. *Building Research & Information*, 23(6), 340−344.

Ellison, L. and Sayce, S. (2007). Assessing sustainability in the existing commercial property stock: Establishing sustainability criteria relevant for the commercial property investment sector. *Property Management*, 25(3), 287−304.

Francis, S., Glanville, R., Noble, A. and Scher, P. (1999). *50 Years of Ideas in Health Care Buildings*. The Nuffield Trust, London.

Freer, N., Lawson, D. and Salter, M. (1999). *Conversion of Redundant Commercial Space to Residential Use*. British Property Federation, London.

Fuster, A., Gibb, A., Austin, S., Beadle, K. and Madden, P. (2009). Adaptable buildings: three non-residential case studies. *Paper presented at the Changing roles: New roles; New challenges conference*, 6-9 October, Rotterdam, The Netherlands.

Gann, D.M. and Barlow, J. (1996). Flexibility in building use: the technical feasibility of converting redundant offices into flats. *Construction Management & Economics*, 14(1), 55–66.

Geraedts, R. and van der Voordt, T. (2002). Offices for living in. An instrument for measuring the potential for transforming offices into homes. *Proceedings of CIB Conference W104, Open Building Implementation*, October 3-4, Mexico City, pp. 207–29.

Gibson, V.A. (1994). Strategic property management: how can local authorities develop a property strategy. *Property Management*, 12(3), 9–14.

Graham, P. (2005). *Design for Adaptability – An Introduction to the Principles and Basic Strategies*. The Australian Institute of Architects, Australia.

Grenadier, S. (1995). The persistence of real estate cycles. *The Journal of Real Estate Finance and Economics*, 10(2), 95–119.

Guertler, P., Pett, J. and Kaplan, Z. (2005). Valuing low energy offices: the essential step for the success of the Energy Performance of Buildings Directive. In *Proceedings of the ECEEE 2005 Summer Study – Energy Savings: What Works & Who Delivers?* 30 May–4 June, Mandelieu La Napoule, France.

Guy, S. (1998). Developing alternatives: energy, offices and the environment. *International Journal of Urban and Regional Research*, 22(2), 264–282.

Hamilton, M., Lim, L. and McCluskey, W. (2006). The changing pattern of commercial lease terms: evidence from Birmingham, London, *Manchester and Belfast. Property Management*, 24(1), 31–46.

Habraken, N. (1998). *The Structure of Ordinary: Form and Control in the Built Environment*. MIT Press, Cambridge.

Hardy, B., Graham, R., Stansall, P., White, A., Harrison, A., Bell, A. and Hutton, L. (2008). *Working Beyond Walls: The Government Workplace as an Agent of Change*. Office of Government Commerce, London.

Heath, T. (2001). Adaptive re-use of offices for residential use. *Cities*, 18(3), 173–184.

Hollis, E. (2009). *The Secret Lives of Buildings*. Portobello Books, London.

Iselin, D. and Lemer, A. (1993). *The Fourth Dimension in Building: Strategies for Minimizing Obsolescence*. National Academy Press, Washington DC.

Jones, P. (2009). A low carbon built environment. *Indoor and Built Environment*, 18(5), 380–381.

Kelly, J., Hunter, K., Shen, G. and Yu, A. (2005). Briefing from a facilities management perspective. *Facilities*, 23(2/8), 356–367.

Kendall, S. (1999). Open building: An approach to sustainable architecture. *Journal of Urban Technology*, 6 (3), 1–16.

Kincaid, D. (2000). Adaptability potentials for buildings and infrastructure in sustainable cities. *Facilities*, 18 (3/4), 155–161.

Kohler, N. and Hassler, U. (2002). The building stock as a research object. *Building Research & Information*, 30(4), 226–236.

Langston, C., Wong, F., Hui, E. and Shen, L. (2008). Strategic assessment of building adaptive reuse opportunities in Hong Kong. *Building and Environment*, 43, 1709–1718.

Lizieri, C. (2003). Occupier requirements in commercial real estate markets. *Urban Studies*, 40(5/6), 1151–1169.

Longcore, T. and Rees, P. (1996). Information technology and downtown restructuring: The case of New York City's financial district. *Urban Geography*, 17(4), 354–372.

Mansfield, J. and Pinder, J. (2008). "Economic" and "functional" obsolescence: Their characteristics and impacts on valuation practice. *Property Management*, 26(3), 191–206.

Mawson, A. (2006). *ICT and Offices: Practiced Realities and Business Benefits*. British Council for Offices, London.

Nutt, B. (1988). The strategic design of buildings. *Long Range Planning*, 21(4), 130–140.

Nutt, B. (1993). The strategic brief. *Facilities*, 11(9), 28–32.

Nutt, B., Walker, B., Holliday, S.,and Sears, D. (1976). *Obsolescence in Housing: Theory and Applications*. Saxon House, Farnborough.

ODPM (2004). *Age of Commercial and Industrial Property Stock: Local Authority Level 2000*. Office of the Deputy Prime Minister, London.

Ohemeng, F.A. and Mole, T. (1996). Value-focused approach to built asset renewal and replacement decisions. In *Proceedings of the Royal Institution of Chartered Surveyors COBRA Conference*. Royal Institution of Chartered Surveyors, London.

Raftery, J. (1991). *Principles of Building Economics: An Introduction.* BSP Professional, Oxford.

Remøy, H. and van der Voordt, T. (2007). A new life: Conversion of vacant office buildings into housing. *Facilities,* 25(3/4), 88–103.

SchmidtIII, R., Eguchi, T., Austin, S. and Gibb, A. (2010). What is the meaning of adaptability in the building industry? *Paper presented at the 16th International Conference of the CIB W104 Open Building Implementation.* 17-19 May, Bilbao, Spain.

Scott Brownrigg, Hilson Moran and Gardiner & Theobald (2009). *Can Do Refurbishment: Commercial Buildings of the 70s, 80s and 90s.* British Council for Offices, London.

Searle, G. (2002). Uncertain legacy: Sydney's Olympic stadiums. *European Planning Studies,* 10(7), 845–860.

Shipley, R., Utz, S. and Parsons, M. (2006). Does adaptive reuse pay? A study of the business of building renovation in Ontario, *Canada. International Journal of Heritage Studies,* 12(6), 505–520.

Silver, N. (1994). *The making of Beaubourg: A building biography of the Centre Pompidou, Paris.* MIT Press, Cambridge, Massachusetts.

Slaughter, S. (2001). Design strategies to increase building flexibility. *Building Research & Information,* 29 (3), 208–217.

Stanhope (1992). *An Assessment of Imposed Loading Needs for Current Commercial Office Buildings in Great Britain.* Stanhope, London.

Stanhope (2001). *A Review of Small Power Provision and Occupational Densities in Office Buildings.* Stanhope, London.

Till, J. (2009). *Architecture Depends.* MIT Press, Cambridge, Massachusetts.

Weatherhead, M. (1999). *Real Estate in Corporate Strategy.* Macmillan, London.

Weeks, J. (1963). Indeterminate architecture. *Transactions of the Bartlett Society,* 2, 85–105.

Weeks, J. (1965). Hospitals for the 1970s. *Medical Care,* 3(4), 197–203.

Wilkinson, S., James, K. and Reed, R. (2009). Using building adaptation to deliver sustainability in Australia. *Structural Survey,* 27(1), 46–61.

Williams, A.M. (1985). *Obsolescence and Re-use: A Study of Multi-storey Industrial Buildings.* School of Land and Building Studies, Leicester Polytechnic.

Worthington, J. (2001). Accommodating change – emerging real estate strategies. *Journal of Corporate Real Estate,* 3(1), 81–95.

Yohanis, Y. and Norton, B. (2002). Life-cycle operational and embodied energy for a generic single-storey office building in the UK. *Energy,* 27, 77–92.

4 The Change Management Challenge in Growth Firms

Paul Dettwiler

CHAPTER OVERVIEW

How are growth firms affected by changing needs? That is, firms that are undergoing a period of expansion and have yet to commit to a facilities solution. Commitment to such a solution potentially provides a 'straight jacket' for future growth. In addressing this dilemma this chapter considers:

1. the necessity to discern relevant needs in order to optimise limited resources,
2. the dynamics of facilities management among growth firms,
3. the multitude of factors from the external environment that impact on the interests of stakeholders and finally
4. an approach for a more effective identification of relevant needs.

This chapter investigates the concept of 'need' and the dynamics of facilities management. Furthermore, the chapter calls into question whether or not a perceived need should prompt change. Defining 'needs' accurately is a prerequisite to meeting the demands for spatial flexibility. This task is particularly pertinent to growth firms for whom the future is most uncertain and the money available for property investment is scarce, in particular growth firms in the start-up phase. 'Needs' in this chapter are regarded as the drivers of change in the spaces and services of users of buildings. However, external factors from the surrounding world ultimately become internal factors and thereby gain relevance for local decision makers; thus 'needs' should not be restricted to users — they should include facilities managers, architects and other stakeholders (owners, investors, etc.). Growth firms, in particular, are subject to the dynamics of interior and exterior forces where professional roles are indistinct. Empirical data from a survey of growth firms, primarily in Sweden (but also with observations from Hong Kong and Italy) provides a basis for a conceptualisation of needs as described in this chapter.

The skill of categorising and discerning 'needs' provides a mechanism for more insightful and efficient managerial decisions, minimising bias and erroneous decision making.

Facilities Change Management. Edited by Edward Finch.
© 2012 Blackwell Publishing Ltd. Published 2012 by Blackwell Publishing Ltd.

Various professional roles that have an impact on the early conceptualisation of needs can be linked to five categories of the external business environment:

1. Market and Financial Forces,
2. Attitudes and Cultures,
3. Politics and Government,
4. Environment and Resources, and
5. Research and Development (influencing both core business and support business).

The concept of 'need' is identified and analysed against four background factors, illustrated within a matrix:

1. Function,
2. Image,
3. Reactiveness and
4. Proactiveness.

Decision making is often based on perceived rather than actual need. The change and change process itself might be regarded as a positive event for the organisation (being seen to be doing something). The challenge for the future is to provide relevant knowledge (supported by ICT technology and knowledge management) to make decisions that are reversible through the embedding of flexible mechanisms in the briefing process. This approach is particularly pertinent to the space provision of growth firms and entrepreneurial firms.

Keywords: Briefing; Contingency theory; Decision; Business environment; Facilities management; Need; Growth firms.

4.1   INTRODUCTION

This chapter relates to the issue of obsolescence discussed in the preceding chapters and how performance decreases over time. Brand (1995) highlighted that all buildings have an intrinsic feature of obsolescence (in positive and negative terms), due to the contrast between the long lives of buildings compared to the continuously changing purpose of spaces. Creating flexible spaces plays a key role in the briefing process in order to stem the onset of obsolescence, both for new constructions and refurbishments. The character of the associated needs thus changes over time. van Marrewijk (2009) stressed the direct links between (1) organisational cultural change and (2) physical arrangements, artefacts and architectural concepts and proposals. Change managers should thus be included at the early stage of the design process because organisational change becomes a physical change through architecture, which firmly connects the briefing process, in which needs should be clarified. The briefing process is largely dependent on weighting issues appropriately: for example, costs must be balanced with quality requirements and users' satisfaction.

This chapter deals with the question: 'When does a need prompt a change in accommodation provision — particularly in relation to the space provision of growth firms and entrepreneurial firms?' The skills involved in accurately identifying needs generate cost-saving benefits; this is because they anticipate and prevent changes that are unnecessary and allow available resources to be allocated to necessary changes in the most appropriate ways.

Change management and organisational transitions have been extensively covered in management literature. However, this coverage rarely focuses on the drivers of change. It can be assumed that a need or a bundle of needs may provide the reason for implementing a change.

There is an underlying assertion in this chapter that an organisation is not a closed entity, but is open and dependent on the external world, which it mirrors. According to contingency theory, factors from the exterior world ultimately permeate the individual organisation as local factors. For example, legislation on ergonomics also satisfies users' expectations and results in higher productivity, leading to higher property value, etc. The different professional roles and responsibilities within entrepreneurial companies tend to overlap each other. Therefore it is unreasonable in this context to assign 'need' to users, facilities managers or stakeholders alone.

Underlying factors of the external business environment are examined that have a causal relationship with the changes in organisations, corporations or companies. Four main questions are considered in the chapter:

1. What are the embryonic change drivers preceding a change initiative that prompts such *changes*?
2. How do external factors influence the conceptualisation of *need*?
3. What factors lie behind the concept of *need*?
4. How are the professional roles of project managers, consultants, architects, facilities managers and owners (clients) related to issues of needs and external factors?

External factors must be collected, analysed, evaluated and put in a local context for the decision maker and can, furthermore, have a greater or lesser element of willingness and enforcement; a company might not choose (or not want) to be affected by external factors (e.g. environmental pollution, wars or other *force majeure* factors have a more compelling character than factors related to the implementation of business strategies or refurbishment, when choices are made between different options).

The facilities management of growth firms is particularly subject to simultaneous interior and exterior forces, as is evidenced in the following study.

4.2 THE DYNAMIC RELATION OF FACILITIES MANAGEMENT VARIABLES AND GROWTH FIRMS

Chalmers University of Technology has in the last two decades performed extensive research on facilities management, entrepreneurship, incubators and growth firms. The study of 'entrepreneurship' confirms the assertion of contingency theory; that is to say, external influences result in complex structures of interdependent needs. A growth firm aiming to conquer a market segment with a service or a product acts in an environment which can be described as hostile because it competes with other market players that do not welcome additional competitors. The founder needs to deal with the problems involved in making decisions within facilities management and property management (e.g. whether to rent, lease or own spaces, choice of location). The facilities management operation of the firm must thus match the conditions of the market and other external factors.

One direction of the research approach at Chalmers University of Technology was the study of growth firms and their simultaneous development of facilities management. Both qualitative and quantitative surveys have highlighted the associated dependence with the dynamics of the prevailing environment. In the initial studies, interviews with CEOs about how they have managed their facilities during the growth of their firms revealed a mix of both appropriate decisions and suboptimal decision-making due to insufficient forecasting of the environment. At the early stages of the growth firm's development, the gap between the spaces and services needed were larger than in later phases. The professional roles that appeared to be distinct in large organisations were mixed and unclear in very young firms: often the founder also functions as the facilities and property manager, user and owner. A three-phase model (comprising entrepreneurial, managerial and consolidated phases) describing the development of the facilities management of growth firms was elaborated by Bröchner and Dettwiler (2004) and Dettwiler and Bröchner (2003). The three-phase model was expressed in terms of key variables enabling a quantitative survey.

One of the observations made concerning the dynamics of space use was that when office space went below $20\,\text{m}^2$ per employee it tended to result in a space change because it was regarded as too congested. This suggested that it was a reactive decision rather than a proactive decision to expand the spaces. Another study of space needs among manufacturing firms in Northern Italy (performed by the institute BEST, Politecnico, Milan) confirms the findings of the Swedish growth firms, namely that space needs are considered by upper management, but in order to reach a decision, a concerted analysis of available alternatives is required as well. In the context of space congestion, which can be accepted to a certain level, other solutions must ultimately be found, such as space extension as well as partial or complete relocation (Ciaramella and Dettwiler, 2010).

A quantitative survey was carried out between 2003 and 2006 that covered all of the 967 firms identified in the Gazelle listing in Sweden, representing all major business sectors. The Gazelle firms appear each year with listings of firms that fulfil stipulated growth criteria regarding growth of turnover and number of employees. It is thus more appropriate to assign the studied firms as *growth* firms rather than start-up firms (they were on average about 10 years old) or incubator firms (only 2% of the population were located in incubators). The survey consisted of a questionnaire of 35 questions relating to two successive three-year periods of low or high GDP. This, in turn, extrapolated to the larger concept of business cycles. Dettwiler (2006) extracted three clusters of growth firms, with the conclusion that various business sectors have different approaches to facilities management. This might be explained by the existence of various corporate cultures and routines for scanning the business environment in order to predict facilities management needs. Dettwiler *et al.* (2006) found that when the business environment becomes more dynamic and hostile, the growth firms avoided investing in real estate, thus favouring rental of spaces, which in turn affected decisions on whether to hire temporary or permanent staff. Firms that act in a hostile and risky environment under an entrepreneurial leadership but are not growing are assigned as *entrepreneurial* firms rather than growth firms. The ability among CEOs to predict the availability of office spaces was slightly higher during recessional periods than during boom periods, which implies an increased ability to evaluate needs when resources are scarce.

Another external factor, cultural differences, tends to influence work patterns and workplace layouts. Dettwiler and Fong (2006) discern significant variations in facilities management between Western and Eastern societies, mainly because working groups and the interaction between them are differently conceptualised, thus impacting on the space layout.

Fluctuations in business cycles (as measures of GDP), for both single and multiple sites (Dettwiler, 2008) have demonstrated significant differences between facilities management variables. Two variables have particular interest in this chapter: firstly the variable 'We were good at predicting our office space needs' which in fact relate to the *management skills* to forecast space needs. Secondly the variable: 'The image of our offices was more important than fulfilling practical needs' which relates to the extent to which functional (or practical) needs were satisfied in contrast to the symbolic values (or the image) (Bröchner and Dettwiler, 2004).

The variables investigated in the quantitative survey do not relate directly to the established profession of 'Facilities Managers' but rather to significant skills relating to FM. It appeared from the results that the roles shifted during the development of the growth firms because the preference for renting or owning spaces changed; spaces were rented during some periods and owned during others.

Some comments on the observations of the Swedish survey follow. Pearson correlations were made of FM-related variables concerning 'The management skills in the prediction of office needs' and the 'significance of the image of offices' (in contrast to fulfilling practical needs).

Observations of the variable 'The management skills in the prediction of office needs' that correlated to FM-related variables used the following assumptions.

1. The preference for renting office spaces appeared negatively correlated during recessions; which might indicate a preference to own spaces (or to take the opportunity to acquire properties at low price during recessions).
2. Staff had earlier experience of space alterations which had influence on their current workplace.
3. The founder's earlier experiences had an impact on the office design.
4. The office (spaces and equipment) supported distance work (during booms but not recessions).
5. The office (spaces and equipment) supported project work in teams.
6. It was important that the office supported individual work needing privacy and offered seclusion and an allotted individual workspace.
7. No correlations of the importance that the spaces enabled informal meetings.
8. That the indoor climate and acoustics were satisfactory in the offices was of no importance.
9. To perform experiments with new approaches to work and workplace design was not important.
10. Layouts and technology (e.g. mobile walls, IT) that made the offices flexible had significance during booms but not recessions.

Observations of the variable 'significance of the image of offices' (in contrast to fulfilling practical needs) that correlated to the same FM variables as above.

1. It was a propensity to rent the office spaces rather than to own them (valid during booms but not recessions).
2. Staff had earlier experience of space alterations which had influence on current workplace.

3. The founder's earlier experiences had *no* impact on the office design.
4. The office (spaces and equipment) supported distance work.
5. The office (spaces and equipment) supported project work in teams.
6. It was *not* important that the office supported individual work needing privacy and offered seclusion and an allotted individual workspace.
7. It was important to be able to have informal meetings inside and outside the offices.
8. The indoor climate and acoustics were satisfactory in the offices was important.
9. They experimented with new approaches to work and workplace design.
10. Layouts and technology (e.g. mobile walls, IT) made the offices flexible were constantly important during booms and recessions.

From the results of quantitative survey we learn that the prediction skills for needs were inconsistent among the decision makers (often the CEOs) in the growth firms. The conception of various FM variables connected to image versus practical needs exhibited both high and low attention, respectively, to certain FM qualities. The next section of this chapter continues towards a generalisation of the problems involved in conceptualising needs and their relation to external factors.

4.3   THE EXTERNAL FACTORS RELEVANT TO FM

For decades a significant role for upper management has been environmental scanning in the support of their core business, where the challenge is to find relevance in information and data. The fact that facilities management units in organisations have grown in strategic importance means that it is also appropriate for those units to pay attention to environmental scanning and enhance or incorporate that activity into their core business. Sensitising start-up organisations to such external changes is particularly problematic, given the uncertainties of the external environment and the need to commit scarce resources.

In the 20th century, companies have applied various approaches to their strategies. The oil crisis of 1973 highlighted severe shortcomings in the planning capabilities of companies which forced the upper management afterwards to focus their strategic work to analyse more on the external world. The historical data of individual firms was shown to be poor at predicting the future, because it mirrored a continuous growth of demand without taking global events into account. Earlier ideas about *contingency* and *systems theory* grew in significance after the oil crisis of 1973. Contingency theory contends that organisations are open systems and must interact with the surrounding world for their survival (Kourteli, 2000). It was then more widely appreciated that organisations are dependent on the external environment. Kourteli contends that 'the external environment within which an organisation chooses to function, determines the internal structure, and overall procedures of the specific organisation', highlighting that collecting large amounts of external data is problematic: it is difficult to find relevance in such data. The mismanagement of data often leads to incorrect decisions on changes being made within organisations. A critique from Kourteli (2000) is that contingency theory is too 'mechanistic' because it overlooks the significance of communication within the organisation.

In current research there is no consensus of how the various factors should be structured when scanning activities of the external environment. In this chapter the external elements that are relevant for facilities management are categorised in terms of five categories expressed as a pentacle, providing the basis of an analytical tool.

Table 4.1   The five background factors for conceptualising needs (relevant for FM).

	Examples of background factors for conceptualising needs (relevant for FM)
Market and financial forces	(1) Variations in supply of competent employees (e.g. old age demographics in Western World) (2) Scope of budget (3) Direct impact of core business dynamics (due to e.g. GDP or market changes) on FM (4) Interest changes affect capability to invest in e.g. refurbishments (5) Opportunities to make leasehold or ownership decisions. (6) Access to corporate real estate, construction land, facility services etc.
Attitudes and cultures	(1) Development and variations of organisational structure and corporate culture (e.g. organisational transparency) (2) Influence on global cultures due to globalisation and demographic changes (3) Development of an egalitarian and tolerant society (4) Organisational democratisation and user co-involvement (e.g. in workplace changes) (5) New work patterns (presence and teleworking).
Politics and government	(1) Legislation on regional and national levels (e.g. sustainability concerning energy), building design (e.g. converted for fire security, disabled people, child security etc.) (2) Standardisation (now common on European Level, e.g. EN norm 15221 concerning FM (3) Decisions on infrastructure, education, health care, army etc. (4) International mobility for work force (5) Decisions to change the repo interest which changes the propensity to invest.
Environment and resources	(1) Finding common interest in environmental measures amongst stakeholders and society (Certification) (2) Energy saving (e.g. through intelligent buildings and electric equipment) (3) Reducing dependence on oil (transfer to alternative energy sources) (4) Use of building material that is environmentally beneficial (e.g. locally produced heavy materials like bricks and stone, or avoiding endangered wood) (5) Planning for flexible buildings enables future activities and avoids prematurely tearing down buildings (e.g. mix of offices and dwellings)
Research and development	(1) Some new technology comes to market early, the lead time after research is short (in fact it is a requirement from some EU-funded programs like Ambient Assisted Living, AAL) (2) Core business paradigms (home care instead of hospital care, emerging technologies, Web, ICT, recycling, replacement of combustion engines) (3) Intelligent buildings (4) Data and information management concerning service and premises.

The five categories, listed in Table 4.1, are interdependent; for example 'having a deliberate interest in taking environmental measures' is relevant in several categories other than 'Environment and Resources'. Perhaps 'Market', 'Attitudes', 'Research' and 'Politics' can also be connected to a considerable extent to taking environmental measures. The nature of interdependence between topics is problematic because it is complex and contradictory; all issues are background factors for change; for example pragmatic questions like: Can we afford changes? Do we need a project? Do we have available competences? What is our regional capacity? How can we achieve competitive advantage? Shall we apply a

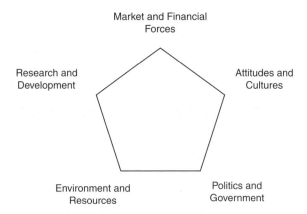

Market and Financial
Forces

Research and
Development

Attitudes and
Cultures

Environment and
Resources

Politics and
Government

Figure 4.1   Structure of the 'Exterior World' where the individual organisation has a minor influence and is subject to the changes.

management style of workers co-involvement? All can be put into the five categories (or if a graphic representation is applied: the five variables and consequently the extremes of the pentacle) of Figure 4.1.

One single factor might have a large impact on the other, for example geographical differences often imply cultural differences; professional roles differ from one country to another. For example an architect in Germany is also active during the construction phase, whereas in Sweden the architect is mostly active in sketching and conceptualisation prior to building permission being gained. The concept of FM is interpreted variably in different countries as well. Dettwiler (2011) proposes a cluster of FM research within the Swiss-German-Austrian region that differs from the Anglo-American region; the former region tends to be more quantitative than the latter, which is more qualitative. In the Anglo-American tradition (Dettwiler, 2011) a significant part of the competence area of a facilities manager is to perform and support changes with the co-involvement of workplace users (Granath, 1999, Blyth and Worthington, 2001). Firms that locate globally have therefore realised the significance of standardising and creating common policies for spaces and services.

In general terms of environmental analysis, the origin of wrong decisions can be connected to inappropriately weighted parameters. It is evident that the limitations of the human mind and time resources make it far from feasible to solve every problem optimally, based on every parameter. For that reason Decision Support Systems (DSS) have been an aid in analytical work during recent decades with widely varying popularity. Interest in DSS has increased, especially in team work, connecting to web technology, data warehouses, OLAP, data mining, web-based DSS and on-line analytical processing (Shim *et al.*, 2002). Often it is in fact a matter of finding and obtaining right data at the right time.

In the FM field, various solutions within CAFM (computer-aided FM) have been proposed. However, IT technology and BIM (building information modelling) are still in its infancy and cannot yet be regarded as a comprehensive and fully trustworthy tool for forecasting. The decision still lies within the professional skills of involved partners. A common problem for facilities managers is to have full acceptance in the upper management of the core business. A main activity of the core business is to scan the environment for the purpose of supporting the core business and the question is whether the proposed environmental scanning should be integrated in core business activities or not. Kourteli (2000) argues for improved communication in order to be successful with environmental scanning.

4.4 EXTERNAL FACTORS RELEVANT TO FM REQUIREMENTS

The complexities of weighing the data and information against each other make it tempting to resolve change decisions by intuition rather than rationality. This might be the reality because even today ICT is not sufficient to solve the problem, and is in fact aimed at the conceptualisation of DSS (decision support systems), project management software and CAFM. The capabilities of CAFM have been questioned by May *et al.* (2007), who see the limits and question the connection of CAD.

Studies with strong co-involvement of users in the design process were made in Sweden in the 1970s, before the ICT era (Olivegren, 1976). The capabilities of information and communication technologies (ICT) of today make it possible to scan the environment (the external world), which was not possible four decades ago. Data can be collected in real time through web technology, whereas collecting historical and future data requires quite different methodologies. Certainly, the future is unknown — therefore different approaches have been used for forecasting: the compilation of historical data, scenario methodologies (e.g. Delphi method), modelling based on known economical or scientific forces (or combined with randomised scenarios).

The increasingly rapid changes make a compelling argument for the supporting services of facilities managers, real estate decision makers and consultants (e.g. architects) to adjust their work according to external changes. The methods of collecting external data and information depend, to a considerable extent, on the perception of the external world. A large problem is the selection of relevant information for a particular situation within an organisation. It is reasonable for the future professionals, responsible for facilities management, to link their scanning of the environment to the existing activities of the core business. Balance must be sought and found to define the areas in which joint and separate efforts should be conducted. One relevant question at the upper managerial level is whether the scanning of the environment should be done separately or conjointly between the core business and the FM unit. Independent scanning carried out by the FM unit might have advantages in avoiding biases and pressure from upper management. Subjective expressions of needs expressed from core business units might be detected and further analysed by the FM unit.

When a 'real' need is identified it can be further decided whether a briefing process with a *pre-project* (Blyth and Worthington, 2001) should be initiated or not. Indeed, a need does not necessarily mean that a change is desirable.

We are all bound to the flow of time and all its events. The human mind perceives time as linear and as a consecutive phenomenon whereby the three main elements (past, present and future) have various influences on an organisation. Forecasting is based on historical data; however, with the emergence of ICT technology and knowledge management in recent decades, the information supplied in real time has given the present more weight in analyses of the surrounding world. The purpose of Figure 4.2 is to illustrate how the time flow of events in the surrounding world might cause a need (or not cause a need). Certainly, the process of identifying needs is as dynamic as suggested by London *et al.* (2005) where the

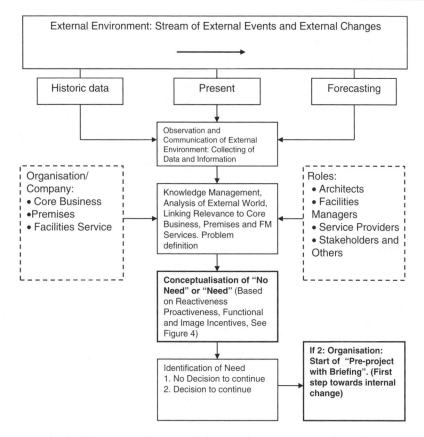

Figure 4.2 The flow from 'external events and changes' towards organisational change.

crucial issue is to overcome barriers by adopting 'reflexive' capabilities (like cross-cultural barriers in international projects, London *et al.*, 2005). Internal communication was also tightly linked to change management by Kitchen and Daly (2002), who emphasise the importance of connecting changes in strategy and organisation to the attitudes among employees.

4.5 DISCERNING THE RELEVANCE OF NEEDS

There are various types of needs; Finch (1993), Blyth and Worthington (2001) and other writers refer to Maslow's 'hierarchy of needs'; however, individual needs must be balanced with organisational needs. They refer, furthermore, to various business sectors that, according to the Orbit study, have 'low change' and 'high change'. (In some circumstances it might even be reasonable to avoid 'high change' because a continuous flow of 'low changes' cause fewer constraints in the organisation; this has been learnt from the pitfalls of outsourcing.)

The needs that result through changes of spaces are often incorporated into a briefing process. Blyth and Worthington (2001) defend the idea of prudence in the very early phases that precede change; the proposed pre-project has a character of reversibility and clarification of alternatives.

Furthermore, in current literature, the word '*gap*' is frequently used to express a distinction between the *status quo* and needs. By identifying real needs with greater reliability it is possible to go a step further than 'gap analysis', whereby significant questions are based on '*what are we today?*', forming the basis of a transition to '*what do we want to be in the future?*' and '*what happens if we do nothing?*' (Internet Source: Office of Government Commerce). Formally, gap analysis considers standards by looking at requirements and standards and finding a way to achieve them. In this context, the idea of a '*gap*' is often used to refer to a shortcoming in the satisfaction of needs. The reason why change does not necessarily satisfy needs might have its roots in the several steps of the early conceptualisation of the need.

As mentioned, a perceived need might be very subjective when not evaluated properly, and might lead to wrong decisions with respect to future changes. Evaluation of changes sometimes reveals that the change decision was wrong or even not at all necessary (Preiser *et al.*, 1988; van der Voordt, 2003). The perception of needs or gaps can be falsely conceptualised. The fact is that the search for a need or a problem can be considered as a bias itself and thus become an artificial construct. For example, architects that participate in architecture competitions often detect that the competition program is filled with unnecessary requirements whereas other necessary topics are lacking. Competition programs within the public sector often have a dominance of political desires that express an image of a new and positive future rather than to focusing on needs that must first be satisfied.Here is another example of the misconception of need:

Buildings in the former DDR Republic from the early 20th century were completely destroyed during the 1990s by very insensitive refurbishments; a great deal of the cultural heritage that had been untouched for a century was replaced by modern building materials that after just 20 years have now faded into obsolescence (observation of the author, active at the time as an architect in Eastern Germany).

What was the 'real need' in this case? The new owners might have felt that a change must be implemented under all circumstances – indeed the facades were dirty (bad image) but were functioning technically – and they conceptualised the 'need' inappropriately (the artificial construct). Perhaps a balance could have been sought favouring more prudent renovation combined with the requirements of our time; however, that was hardly realistic. During a 'boom' in construction, all work is extremely intense which makes it impossible to search for 'prudent' solutions; concepts of time and profitability provide little room for unconventional solutions in the construction sector. Here again, ICT with suitable information will have its role in the future. (Issues of cultural heritage appear later in Chapter 12.)

Managerial attitude to change is a significant topic; Santos Paulino (2009) found that if organisations exhibit *inertia* in risky environments, it is in fact an advantage and there is less risk of failure. It might thus make sense to introduce the notion of *prudence* as a significant element in decision making, or to at least assign it an appropriate weighting in the briefing process.

The complexity of the concept of *need* argues for discovering the underlying factors related to the topic. We can assume that some needs are artificially created because the change itself is desired without any apparent 'real need'. Greiner (1998) highlighted various phases in the lifecycle of a firm's growth and discovered that a change of CEO often leads to organisational changes as well, even if they are not motivated or necessary. For that reason, one can assume *functional* or perhaps rational reasons for conceptualising needs. In contrast, we have irrational reasons that relate to tactical or *image* concerns for an organisation. (This contrast also became a part of the theoretical framework of the survey described in Section 4.2.) However, all factors relating to image need not be negative or irrational. Marketing measures, for example, are in fact image related and a necessary part of business life in order to support the production goals of the core business.

According the European Standard of Facilities Management (EN 15221–1 to 6, 2009), the discipline is divided into three fields: *strategic, tactical* and *operational* FM. To comprehend the intermediary notion, the tactical level, we must concern ourselves with matters such as:

1. Introduction and maintenance of the strategy,
2. Development of the mission and budgets,
3. Translating strategy to operational FM,
4. Management and supervision of regulations,
5. Optimisation of resources,
6. Leading the FM team,
7. Communication performance factors on the tactical level and
8. Adjustments for changes and documentation.

The tactical aspect of needs is thus not to be neglected because this might be where the real needs are detected (being the bridge between operation and strategy); the word *tactical* is contradictory because it is simultaneously associated with something that is done to satisfy image.

Problems often become apparent in retrospect, but future problems are far more difficult to predict. Similarly, the ideas of collecting information mentioned earlier; we have seen that it can be collected at three various parts of the time scale (Figure 4.2): historical data, present or forecasting. Certainly information of forecasting is the result of analysis of historical data, modelling, etc. Inspired by observations and evaluation of decision making in entrepreneurial firms, it seems appropriate to relate the discussion of need to *proactive* and *reactive* behaviour as well as satisfaction of needs as a variable of *image* and *function* (see Section 4.2). The notion of *proactive* must not necessarily be positive, as much as *reactive* must not necessarily be negative. Both approaches of *proactiveness* and *reactiveness* can be well motivated in various situations; however, a self awareness of managers of which of the two approaches is chosen might have significance on the strategy work. We have seen that erroneous forecasting can be disastrous, whereas *reactiveness* can be related to prudent and judicious management. The challenge for the future is to provide relevant knowledge to make decisions that to a greater extent will be reversible through the embedding of flexible mechanisms in the briefing process.

A model is proposed in Table 4.2 that describes:

1. operational FM as functional and reactive fields,
2. tactical FM as predominantly image related and
3. strategic FM as proactive.

Table 4.2 illustrates that each of the four fields of the matrix can house positive as well as negative elements. Furthermore, the three classes of FM (strategic, tactical and operational) are also assigned to the four fields of the matrix. Similarly to a SWOT analysis, the proposed model can be regarded as an analytical instrument that discerns the need: 'What are the

Table 4.2   The nature of underlying variables of needs in FM.

	REACTIVENESS	**PROACTIVENESS**
FUNCTION	**Negative:** Slow reaction due to inadequate communication with upper management and facilities manager.	**Negative:** Too strong reliance on quantitative data (or qualitative) that will steer the identification of needs. Physical data and cost data are not enough in the briefing process.
	Positive: Being 'second' and learning from the mistakes of others or taking advantage of progress in research and development. Standardised and selected FM variables might be appropriate.	**Positive:** Selection of relevant information and analysis of the external world creates the basis for identifying appropriate needs concerning FM.
	Level of facilities management: (1) Operational FM, (2) Services related rather than premises	**Level of facilities management:** (1) Strategic FM, (2) Balance service and premises
IMAGE	**Negative:** Following the mainstream of markets, cultures and attitudes will be a necessary evil if the organisation cannot interpret it appropriately. Corporate culture and image are not consistent.	**Negative:** New leadership wants new management culture with change as a purpose of itself.
	Positive: Implementation of the learning organisation that observes the surrounding world and adapts its needs accordingly.	**Positive:** Well founded changes due to new leadership provide renewal and efficiency gains for the organisation. Corporate culture and Image are consistent.
	Level of facilities management: (1) Tactical FM, (2) Premises-related rather than services related	**Level of facilities management:** (1) Tactical FM, (2) Premises-related rather than services

origins of perceiving this topic as a need?' 'Can we accept a need or must we redefine it?', etc. A primary challenge of scanning the environment is to find and weight relevant information. The proposed model in Table 4.2 would function as a filter and a categorising tool for structuring needs. It is thus not always the case that the need is particularly relevant to the individual end-user, but rather an invented motivation (or excuse) for the organisation to implement a change, because the change and change process itself might be regarded as a positive event for the organisation.

From the survey of growth firms in Section 4.2 it was observed that professional functions were unclearly defined and did not have distinct boundaries. This is also mirrored in larger firms that have a multidisciplinary character. How, then, should different professions be involved in changes of spaces and services related to the external factors of Table 4.1? The upper management (representing the users) of a corporation is totally responsible for the whole organisation, both core and support, which means all five fields ought to be covered in an environmental scan; in particular 'Research and Development' topics for developing the core business are not covered by professions related to property and FM. Individual project managers (both within the core and support business) have a coordinating function and should in fact not be specialists in any of the five fields shown in Table 4.1. Facilities managers and architects are expected to possess knowledge especially within the fields of 'Attitudes and Cultures' and 'Environment and Resources'. In addition, architects should be updated with legal regulations, whereby 'Politics and Government' is their field of competence as well. Certainly more external fields of environmental scanning are relevant, depending on the activity level of FM (operational, tactical and strategic). The client (the owner, the contractor, the investor of premises) has a similar interest to upper management in scanning all external factors, whereby common targets and partnering become relevant.

Managerial awareness of the background of the needs described is useful for the conceptualisation of needs. The categories of the external environment must be related to professional competences. Certainly a property manager, for example one that acquires and sells property, has more interest and competence in the fluctuations of interest rates than an architect, whereas an architect pays closer attention to new legislation relevant for a design project than the owner.

Skills for executing change have become a significant component of the basic skills of management consultants. Worren *et al.* (1999) regard change management as the service of major consulting firms and compare change management with traditional organisational development (OD), in terms of corresponding theory and analytical framework. Worren *et al.* (1999) adopted the word 'facilitator' to describe change management specialists. In the creation and writing of briefs, Kelly *et al.* (2005) forecasted new professional roles for facilities managers.

4.6   SUMMARY

This chapter has explored the concept of need and the very origin of the *raison d'être* of changes, and has also explored the philosophical question 'What is the real need behind a need?' A generalisation is made through studies of the dynamic conditions in which growth firms have to make decision of the future space needs. The organisation is regarded as an exogenous system where the surrounding world sooner or later permeates into the organisation and affects the various professions that are involved in the briefing process. Rationality is sought in the massive flow of information from the external world that influences decision makers. A reinforced role of ICT and DSS in the decision process is argued for because the objective is to achieve a higher degree of reversibility in decisions that cause changes in buildings and services. A four-field matrix is presented with the scales image and *function* versus *reactiveness* and *proactiveness* as an aid to clarify thoughts before and during the very early stages of the briefing process. Enabling the extraction of the 'real needs' can improve efficiency and, above all, avoid expensive decisions in future briefing processes. Due to the inter-disciplinary and coordinating skills of facilities managers, they will have an opportunity to take on the professional role of identifying the real needs before change occurs, and to play a key role in the pre-briefing and briefing stages of a construction project.

REFERENCES

Blyth, A. and Worthington, J. (2001). *Managing the Brief for Better Design.* Spon, London.
Brand, S. (1995). *How Buildings Learn: What Happens After They're Built.* Penguin Books Publisher, New York.

Bröchner, J. and Dettwiler, P. (2004). Space use among growth companies: linking the theories. *Facilities Management: Innovation and Performance*. Ed. Alexander, K., Atkin, B., Bröchner, J. and Haugen, T. London: Spon Press, pp. 47–58.

Ciaramella, A. and Dettwiler, P. (2010). Influence of FM factors on location decisions of manufacturing firms. In *Proceedings CIB W070 International Conference in Facilities Management, FM in the Experience Economy*, São Paulo, Brazil, Sept. 13–15 2010, pp. 383–394.

Dettwiler, P. (2006). Offices of Swedish growth firms: facilities management variables. *Facilities*, 24(5/6), 221–241.

Dettwiler, P. (2008). Modelling the relationship between business cycles and office location: The growth firms. *Facilities*, 26(3/4), 157–172.

Dettwiler, P. (2011). Characteristics of real estate and facilities management issues in Switzerland: Knowledge transfer between business life and academia. In *Evolution of the Process of Building Production in the European Context*, Chapter 5.1.3. Politecnico di Milano.

Dettwiler, P. and Bröchner, J. (2003). Office space change in six Swedish growth firms. *Facilities*, 21(3/4), 58–65.

Dettwiler, P. and Fong, P.S.W. (2006). Learning from Sweden: transfer of facilities management related variables of growth firms to China. *Proceedings, CIB W70 Trondheim International Symposium Changing User Demands on Buildings*, 12–14 June, 2006, pp. 611–620.

Dettwiler, P., Lindelöf, P. and Löfsten, H. (2006). Business environment and property management issues: a study of growth firms in Sweden. *Journal of Corporate Real Estate*, 8(3), 120–133.

European Standard of Facilities Management, EN 15221-1 to 6 (2009), CEN, Comité Européen de Normalisation, CEN/TC348.

Finch, E. (1993). Facilities management at the crossroads. *Property Management*, 10(3), 196–205.

Granath, J.Å. (1999). Workplace making: A strategic activity. *Journal of Corporate Real Estate*, 1(2), 141–153.

Greiner, L.E. (1998). Evolution and revolution as organizations grow. *Harvard Business Review*, 76(3), 55–67; 1972 version is in Vol. 50(4), 37–46.

Kelly, J., Hunter, K., Shen, G. and Yu, A. (2005). Briefing from a facilities management perspective. *Facilities*, 23(7/8), 356–367.

Kitchen, P.J. and Daly, F. (2002). Internal communication during change management. *Corporate Communications: An International Journal*, 7(1), 46–53.

Kourteli, L. (2000). Scanning the business environment: some conceptual issues. *Benchmarking: An International Journal*, 7(5), 406–413.

London, K., Chen, J. and Bavinton, N. (2005). Adopting reflexive capability in international briefing. *Facilities*, 23(7/8), 295–318.

May, M., Marchionini, M. and Schlundt, M. (2007). CAFM – Status Quo. In *FM Praxis*, Bauverlag BV GmbH, pp. 5–34.

Olivegren, J. (1976). http://libris.kb.se/preferences.jsp?tab=gen *Brukarplanering: ett litet samhälle föds: hur 12 hushåll i Göteborg planerade sitt område och sina hus i kvarteret Klostermuren på Hisingen*. Dissertation, Kungliga Tekniska Högskolan, Stockholm.

Preiser, W., Rabinowitz, H.Z. and White, E.T. (1988). *Post-occupancy evaluation*. Van Nostrand Reinhold, New York.

Santos Paulino, V.D. (2009). Organizational change in risky environments: space activities. *Journal of Organizational Change*, 22(3), 257–274.

Shim, J.P., Warkentin, M., Courtney, J.F., Power, D.J., Sharda, R. and Carlsson, C. (2002). Past, present, and future of decision support technology. *Decision Support Systems*, 33, 111–126.

van der Voordt, D.J.M. (2003). *Costs and Benefits of Innovative Workplace Design*. Center for People and Buildings, Delft.

van Marrewijk, A.H. (2009). Corporate headquarters as physical embodiments of organisational change. *Journal of Organizational Change Management*, 22(3), 290–306.

Worren, N.A.M., Ruddle K. and Moore K. (1999). From organizational development to change management: The emergence of a new profession. *The Journal of Applied Behavioral Science*, 35(3), 273–286.

5 The Business of Space

Danny Shiem Shin Then

CHAPTER OVERVIEW

This chapter provides an overview of space planning from a business-driven perspective. The focus will be on how to plan for and manage facilities change resulting from office reconfiguration and office moves by regarding functional spaces within buildings as a business resource that must be optimised – hence 'the business of space' (McGregor and Then, 2001). The provision of functional workspace is no longer simply determined by financial considerations. Demand assessment at a corporate level is multi-dimensional: it therefore needs to take into consideration location preferences and transportation convenience, technological developments that support flexibility and mobility, corporate image and culture, and individual tasks and preferences. Clearly, facilities change management is inextricably linked to the provision and management of business facilities in which the workplace and workspace environment is the embodiment of the corporate culture and branding.

Understanding the status quo and projecting future requirements are key ingredients in effective space planning and management – understanding the baseline position. The continuing pace of developments in computing and communication technologies, increased mobility and social networking tools will mean that there are many possible alternatives to the traditional workplace or work style. The design and provision of a correctly balanced infrastructure of physical and virtual workspaces will be at the heart of strategic space planning for modern corporations.

This chapter is structured to provide a systematic assessment of demand and supply variables that impact on decisions relating to the provision and utilisation of functional space as a business resource. The importance of having a coherent corporate accommodation strategy that drives decision making is paramount in managing space demand over time. The business of space has resulted in a new geography of workplace that calls for careful evaluation of real estate portfolio options.

The design and reconfiguration of space within organisations, whether in terms of physical relocation or workplace reconfiguration, involves not only physical changes but also emotional changes for the people affected. Effective facilities change management will require a thorough process of assessing organisational needs, assessing supply and deriving an appropriate outcome by matching demand and supply. Success in such a process will rely on taking a vision of the future and a structured approach in reaching it. This, in turn, needs to be supported by a comprehensive understanding of the business

Facilities Change Management. Edited by Edward Finch.
© 2012 Blackwell Publishing Ltd. Published 2012 by Blackwell Publishing Ltd.

ds, data from the organisation and physical characteristics of the building infrastructure (including constraints). In space planning and management, the focus must be on the people using the facilities, the technology needed to support tasks and work patterns, and the management skills, together with necessary systems capabilities to adjust and monitor facilities performance.

Keywords: Business of space; Technology and workspace; Workplace strategies; Space planning and management.

5.1   INTRODUCTION

In today's business environment the business support infrastructure are no longer static. In terms of physical asset infrastructure, company directors are increasingly forced to reassess their work environment — not only their location, size and internal configuration, but the fundamental economics of business-resource utilisation. The nature of work, ways of working and work environments are in a state of flux as we attempt to adapt to meet continuous change in global market trends and economies. Consequently, in order to ensure that the most fitting work environments are provided to meet the needs of business units, company directors tasked with the responsibility of developing workplace strategies and planning workspace must constantly consider their future needs. The workplaces of the future must not only accommodate rapid changes — of a political, economic, technological and social nature (Oseland, 2008) — but must also strive to reflect and promote new ways of working.

The proactive alignment of real estate assets and workplace strategies with corporate goals and objectives of business units will continue to be the push for company directors (Osgood, 2009). Strategies aimed at enabling the business will be of core concern for real estate and facility management. At the same time, the new workplace must take on adjectives like *global, virtual, responsive, agile, adaptive* and *flexible*, in order to cater for a multi-generational and diverse workforce. Successful implementation of appropriate workplace solutions will demand a thorough understanding of the drivers of business occupancy costs and how they can be systematically assessed, evaluated, procured and managed to remain strategically and operationally relevant in supporting the current corporate strategy. Success begins with strategic planning supported by integrated, robust processes (Acoba and Foster, 2003; Then, 2003a; Haynes and Nunnington, 2010).

5.1.1   Space as a business resource

Space as a neglected business resource is perhaps more in evidence in recent years as a result of the global financial crisis which has presented many companies with surplus space as a consequent of strategies to reduce headcount, and such companies have come to realise the significance of premises costs associated with occupied space.

The general push of most real estate/facility management strategy focuses on enabling the business, helping to improve business productivity, managing costs and providing business process support. This push is set against a dynamic business environment that demands a degree of flexibility via multi-year planning and multiple scenarios for growth or contraction. While factors like air quality, views and natural lighting are important, other factors like corporate branding, demographic and diversity of workforce are dimensions that can no longer be ignored in space planning and management.

5.1.2   Technology and its impact on the corporate workplace

Globalisation and technology together are creating the potential for startling changes in how we do our jobs and the premises/location we do them in. The modern workplace no longer resembles the factory assembly line but rather the design studio, where the core values are collaboration and innovation, not mindless repetition. The role of information and accessibility to information are critical components of today's increasingly 'distributed' and yet 'connected' work environment. The age of the social networking, internet and wireless accessibility is upon us — work is becoming more collaborative and team-based (Becker and Steele, 1994; Grantham, 2000; Cairncross, 2001; Froggatt, 2001; Malone, 2004).

A recent report (Perske *et al.*, 2009) makes the case that mobility is inevitable and is the only way companies around the world will be able to remain competitive and financially and environmentally viable. Technology driven capabilities that impact work and workplace design will continue to evolve. Given the diversity of businesses and corporate cultures, it is apparent that no one size will fit all. The resulting corporate workplace will be a resolution of an amalgam of factors:

- nature of business and business delivery processes
- nature of work and work-enabled tools and systems
- corporate culture and its values
- corporate vision of its future.

5.2   CONTEXT OF SPACE PLANNING AND MANAGEMENT

Space planning and management is about anticipating and managing facilities change to align with corporate strategy direction. It is difficult to predict future facility needs and develop strategies that will enable a timely response. Spatial need is a good example of the importance of strategic management of corporate facilities. Most organisations will require suitable workspace for its staff, but the amount and types of space will grow and shrink over time as a result of economic conditions, business performance, new initiatives and changes in technology. The pace of change has meant that facilities change is both inevitable and necessary. The drivers for change in terms of space planning and management are no longer solely financial but may involve other 'soft' corporate motivations that are related to corporate culture and branding, and attracting talents.

5.2.1   Business management and economic drivers

It is crucial for corporate directors to understand the resource implications of corporate real estate assets in terms of:

- the importance of the workplace environment as an enabling facility that shapes corporate culture and behaviour
- the economics of the provision of real estate assets as operating resources, as well as in terms of their intrinsic value
- the relationship between the physical environment, individual satisfaction and organisational productivity
- the provision and ongoing management of facilities support services and their users interface with the workplace environment.

Figure 5.1   Business drivers and affordability drivers (Then, 2005).

The real estate and facilities management functions, for their part, have to move from being a transactional-reactive role to a strategic-proactive role, where the emphasis is on anticipating the future in the light of the company's core business and its work processes, whilst consistently providing value-adding solutions. Enabling business means increasing productivity. In this respect there is a greater awareness that in any given building, 90% of the operational costs are people driven with 10% property driven. A 2 or 3% improvement in productivity has a much greater impact then cutting 5 or 10% of property costs (McCann and Venable, 2010).

The critical interfaces between *facilities provision* and *facilities services management* within an organisational context are illustrated in Figure 5.1. The objective is to identify the crucial interfaces and to identify the drivers motivating actions at three identifiable levels, i.e. corporate level, estate level and building level.

The *corporate level* is concerned with the adequacy of the real estate assets, as a business resource, in fulfilling strategic objectives. The *estate level* then interprets this strategic intent in terms of implications for the current operational real estate portfolio, i.e. facilities provision. The *building level* is primarily concerned with meeting users' requirements on an on-going basis. This needs to be achieved with minimal disruption while action is taken to adjust to the next 'steady state'. It is important to point out that underpinning any decisions in facilities provision is the constant interplay between the pulls from three key resource drivers: those of people, technology and the workplace environment.

5.2.2   Business planning and space planning

In the business world there is growing acceptance of the suggestion that, in order to achieve the much-needed alignment between business strategic direction, organisational structure, work processes and the enabling physical environment, an organisation's strategic intent must clearly reflect the facilities dimension in its strategic business plans. Both published literature and practice evidence (Wilson, 1991; Nourse and Roulac, 1993; Joroff *et al.*, 1993; Graham Bannock and Partners Ltd., 1994; Then, 1994, 1999; Arthur Anderson, 1995; Avis and Gibson, 1995; Apgar IV and Bell, 1995; Gallup, 1996; The Henley Centre, 1996) point to three prerequisites for a strategic approach to the effective management of workplace provision and management:

- the need to link real estate/facilities decisions to corporate strategy
- the need to proactively manage functional workspace as a business resource and
- the need for a framework to integrate business resource management to the provision and management of the corporate operational assets and associated facilities support services, in their business settings.

In providing facilities solutions to meet emerging business challenges it is becoming increasingly important to show the link between work and the workplace, between workplace provision and the effectiveness of people. An important correlation is that between the way the workplace is configured and serviced, and businesses' effectiveness in meeting the needs of their customers.

One of the key responsibilities of facilities managers in managing demand over time is to ensure that the original assumptions that formed the basis of the accommodation strategy are still valid, and remain so, set against the ever-changing business environment. All too often facilities management is reduced to maintenance, engineering or office services, but space management is a strategic role. Used well, space is not purely a cost, but a generator of revenue in its own right. Well-designed and managed space motivates staff as well as attracting and retaining customers. Any review of an accommodation strategy today must consider the implications of changes in the business, changes to the work practices and changes to workspace; on the existing operational portfolio and the likely future need for other buildings. Hence, the provision of workspace is a strategic issue and the process of maintaining the strategic relevance of the accommodation strategy is a continuous one.

5.3   STRATEGIC SPACE PLANNING — THE ACCOMMODATION STRATEGY

The planning of all workspace should be a response to specific business needs and corporate drivers. The challenge of managing workspace is to achieve the best match of the supply of functional space to the dynamics of changing business demand. The underlying business rationale of effective workplace management is the promotion of business effectiveness. The effective management of workspace is based upon two imperatives — firstly, the acceptance of the economic reality that every square metre of occupied space has to be paid for by the business; and secondly, that the space occupied by staff should be directly related to the requirements of the work processes and tasks the person has to perform, rather than based on their seniority and status. It is from this starting point that the relationship between business planning and space planning becomes clear. The process of addressing the workspace needs of the business requires the conscious planning and design of the work environment to facilitate the delivery of products and services, driven by desired business practices (culture and organisation) and operational requirements (tasks and functions).

Just as the aim of a business strategy is to provide a framework which guides the organisation in its decision making towards a chosen goal, so too an accommodation strategy provides the framework for decision making relating to the management of an organisation's workspace. Unfortunately, all too often the fact that an organisation evolves rapidly, whether it is by organic means or by acquisitions and mergers, is seen as justification for not producing a strategic accommodation plan. The results all too often experienced as a direct consequences of the failure to plan result in the achievement of poorer than expected productivity levels. These typically include inappropriate premises, insufficient available workspace, inefficient facilities and services, outmoded systems for handling information and conflict between facilities providers and facilities users, resulting in poor staff morale. In order to ensure success, there is a need to consider workplace-related issues at an early stage, and in so doing, create the framework around which the organisation's workspace can be developed to function effectively. This is where the need for a *strategic accommodation plan* becomes critical (Figure 5.2).

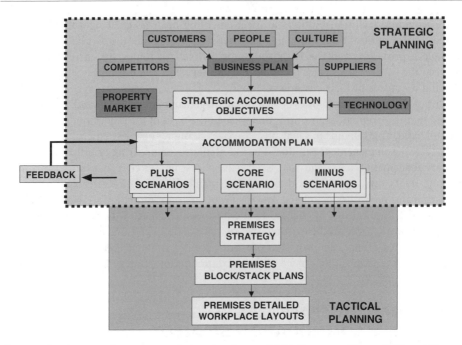

Figure 5.2 Elements of an accommodation planning model (McGregor and Then, 2001, p. 37).

The production of an accommodation plan fosters an atmosphere in which decisions relating to the business's workspace can be made, confident in the knowledge of being able to respond to the varying needs of the organisation through time. The accommodation plan has its roots in the organisation's business plan and is itself a component of the organisation's facilities plan. When well constructed, the plan not only enables *responsive actions* to be initiated, but will provide the basis for *anticipatory actions* by the real estate/facilities manager. This can be said to be the challenge for all workspace planning — to move from a *reactive* to *anticipatory* process.

Effective strategic facility planning requires constant and continuing communication between a business unit's planners who chart the future path for the company and the real estate/facilities professionals who are charged with delivering the physical infrastructure to affect the business plans. The use of *scenario planning* provides a useful mechanism of promoting the much-needed dialogue between strategic planning and tactical planning. It also provide for the two-way exchange between *business* information and *building-facilities* information. This informed interface between corporate planners and real estate/facilities managers should be driven by a clear understanding of the following business demand variables:

- the push to control costs and minimise long-term commitments to infrastructure — both of which suggest the removal of less functional space
- the increasing need to provide workplaces that enhance productivity, while addressing increasingly complex environmental and technological requirements
- the provision of a satisfactory work environment for employees, individually and collectively.

In far too many instances, organisations have suffered through the failure to plan effectively. Effective plans can be developed by means of scenario planning. Prasow and Sargent (2009) emphasised that 'employing scenario planning continuously to maximise the organisation's ability to adjust and react will set well-prepared companies apart from those who will continuously be scrambling.'

Proper evaluation of scenario options can save much corporate pain, both physical and financial, in the future; as this is where the linkages between the business and accommodation plans are made and where strategic and tactical plans play a vital role in the successful application of the accommodation forecasts. It is for this reason that the real estate/facilities manager should always be working to a 5-year planning horizon, using various scenarios to take account of changes that may arise in the intervening period.

From the information gathered, a *core scenario* can be developed which is the organisation's best view of the future. 'Best' in this context is not the 'most optimistic' but is the 'best forecast' based on a full understanding of the business environment in which the organisation operates. From this core scenario, other possible scenarios are developed, by applying sensitivity analysis techniques. From these 'alternative worlds' of business, several corresponding accommodation scenarios can be developed to meet the needs of different levels of business, different numbers of people in the organisation, different levels of technology usage, different organisation structures and so on (McGregor and Then, 2001, Chapter 3).

5.4   ASSESSING DEMAND — ORGANISATIONAL NEEDS

The planning of all workspace should be a response to specific business needs. A key role of the facilities manager is the interpretation of business data into a set of requirements for workspace and its infrastructure - translating the 'business speak' of managers into facilities language, to enable possible solutions to be developed, and then vice versa, to present workspace proposals for consideration by the business. This process is data intensive and is structured to gather information about the current and future needs of the business, and the people within it.

The level of detailed information required will vary depending upon the level of the planning process, i.e. strategic or tactical. Also, in certain circumstances, some processes and techniques to gather the data are more appropriate than others. It would, however, be unrealistic to expect that business managers, unless they are unusually skilled in this area, to be able to provide the information that the facilities manager can use directly to plan the organisation's workspace requirements. Therefore, however it is gathered, the information required by the facilities manager will pass through a process of conversion from business information into workspace data (see Figure 5.3).

Typically the types of information the facilities manager will require to enable planning of workspace to be carried out will be grouped under people, workspace and services, as listed in Table 5.1.

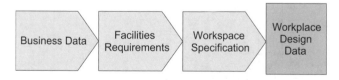

Figure 5.3   The process of converting business data into workplace design data.

Table 5.1 Information requirements for assessing demand

People	Workspace	Services
Types and numbers of people using the building and its facilities including other parties who are visitors for short and long duration. Number of people to be engaged in each of the work processes of the business. Anticipated occupancy of the premises in terms of number of people, timing and duration.	Types and numbers of people using the building and its facilities including: • types of workspace required and its attributes • types of workspace settings required, e.g. individual and collaborative workplaces, meetings, social, open and enclosed. Internal locations of workplaces – aggregated or dispersed. External locations of workplaces – CBD premises, satellite premises, customers and supplier premises, transit locations and staff homes. Durations for which the various workspaces are required and by which workgroup. Process operating environment and conditions, e.g. segregation, co-location.	IT infrastructure: required to support the business, e.g. IT and communication systems, LANs and WANs, telephones, etc. Support services: required for each work group and the business as a whole, such as security, reprographic, catering, cleaning, etc. Operating hours: the hours of operation of the business and its constituent parts. Equipment and machinery. Information: types of information required to support business processes and its anticipated pattern of use. Storage: e.g. document, records, raw materials, finished goods and consumables, in support of work processes and their anticipated pattern of use.

The above data groups are assessed at work group level – and by aggregation of the needs of each group at the organisation level. The outcome of demand assessment is a statement of the workspace requirements for the organisation as a whole, the synthesis of which is used to develop an accommodation plan. Figure 5.4 illustrates elements of workplace design data.

In gathering the data, whether at strategic or tactical levels, it is important to distinguish between the *wants* and *needs* of the people in the business. The use of appropriate data-gathering tools can go a long way to assisting the facilities manager with the screening out of undesirable 'wish lists'.

A major role of facilities management is the efficient operation of all serviced spaces supporting the delivery of the core business activities. Operational support on an ongoing basis demands a thorough understanding of work tasks carried out, and the criticality of the processes supporting business delivery. Ongoing support tasks can be grouped under two categories of activities.

- Measures to ensure the smooth operation of serviced facilities in terms of:
  - the range of building-related services – utilities and statutory requirements, building maintenance and repairs
  - the range of business support services – cleaning, catering, office services, etc.
- Measures to handle short-term fluctuations of demand for workspace – churn management.

Whilst the former functions are essentially transaction oriented, with the focus on meeting agreed service levels, the latter are more strategically driven by organisation culture and vision of 'how we want to work'. The perceived cultural inclination and strategic visioning will be informed by key variables such as:

- categories of workplaces – enclosure-based, ownership-based, activity-based and time-based
- design of workspace – variety in work settings (both off and on-premises), furniture selection and provision for flexibility
- protocol to support implementation of alternative workplace practices
- provision and management of essential technical support infrastructure – technology and communication
- changing demand variables – impact of remote and distributed working, multi-generational workforce, work-life balance, social responsibilities and sustainability, etc.

5.5   ASSESSING SUPPLY — PREMISES AUDIT

The premises audit is the supply side of the space planning equation. The supply variables are concerned with factors that influence decisions about the provision of real estate resource and the servicing of workspace in support of business processes. The objective of a premises audit is to assess how well a building is currently meeting the needs of the occupying business, and the degree to which it could satisfy their future needs. Many of the issues are similar to those that would be required as part of the process of evaluating new premises. The key issues of a premises audit will encompass location, condition, utilisation and value for money.

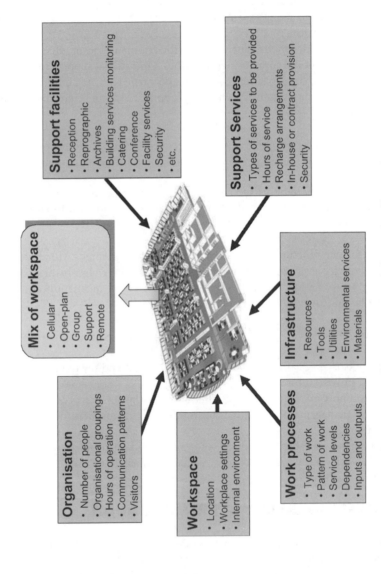

Support facilities
- Reception
- Reprographic
- Archives
- Building services monitoring
- Catering
- Conference
- Facility services
- Security
- etc.

Support Services
- Types of services to be provided
- Hours of service
- Recharge arrangements
- In-house or contract provision
- Security

Mix of workspace
- Cellular
- Open-plan
- Group
- Support
- Remote

Infrastructure
- Resources
- Tools
- Utilities
- Environmental services
- Materials

Organisation
- Number of people
- Organisational groupings
- Hours of operation
- Communication patterns
- Visitors

Workspace
- Location
- Workplace settings
- Internal environment

Work processes
- Type of work
- Pattern of work
- Service levels
- Dependencies
- Inputs and outputs

Figure 5.4 Elements of workplace design data (summarised from McGregor and Then, 2001, Chapter 4).

Effective utilisation of the premises is determined by the suitability of the environment for the work tasks to be performed. Therefore, in constructing procedures for the conduct of premises audits, it is essential that the needs of the business are known as they will act as the backdrop against which the evaluation process will be conducted. Consequently the audit will, in addition to assessing the condition of the fabric and services of the building, also take account of how the premises are used. Analysis of the audit data will enable the facilities manager to identify aspects requiring attention, such as the maintenance of the building's fabric, improvement of services and the optimisation of workspace layouts.

Clearly the single most important aspect of the premises audit is to identify priority areas, to ensure that the occupier receives the best value for money from the premises. This will include assessments of energy efficiency, adaptability and cost in use, bringing together all of the strands of the audit, i.e. location, condition and utilisation. Some 'cost in use' studies may be more easily carried out than others. For example, an assessment of the premises energy running costs would not be too difficult to undertake, drawing the information from fuel bills and occupancy periods. Assessing value for money in adaptation and internal changes may be more difficult to achieve. However, in organisations with high levels of churn, the costs associated with premises adaptation may have a significant impact upon facilities expenditure. The audit should therefore include an assessment of the costs of implementing workplace changes, be they partition moves, data/communications cabling re-configurations or others, and assess them against the provisions made for flexibility, i.e. the provision of, and investment in, re-locatable partitions, raised floors and the like.

As with all audits, the premises audit only has a value to the organisation if the investigations are applied as learning and feedback for the benefit of future premises operations. Typically we would expect a full audit, such as that described here, to be undertaken every five years. However, it is likely that some aspects, such as a review of the utilisation of the premises, may be undertaken more frequently, possibly on an annual basis where an ongoing programme of churn and subsequent post-occupancy evaluation can be put in place.

To summarise, in assessing the supply side of the business of space, the functionality and quality of facilities are at least as important as the amount of space to be occupied. For facilities managers, building and space audits can be extremely useful in developing a thorough understanding of:

- quantitative information on the existing building stock, and other premises under consideration, and affords the opportunity to revisit the premises policy
- qualitative information on the appropriateness of the buildings in meeting current and projected demands for workspace, and an opportunity to test actual space usage against an effective space budget and planning concept
- financial options for maintaining or modifying existing buildings, or acquiring new buildings to meet projected demand
- the need for the development of an appropriate methodology and corporate guidelines, for use in making a smooth transition from a traditional supply-led provision of workspace, to an appropriately designed demand-led work environment.

5.6   RECONCILING DEMAND AND SUPPLY — FACILITIES SOLUTIONS

Successful space planning resolves *supply* and *demand* based upon an understanding of the present use of space, and in doing so establishes a profile for the future use. It is important to understand space, its significance as a strategic asset as well as an operational resource, and plan for its change and adaptation over time. Perhaps, even more significant, is senior management's endorsement that the working environment can be used as a catalyst for change. The challenge is to effectively match supply to demand for space over time guided by an agreed strategic framework and an objective demand-led analysis of workspace.

In promoting a strategy-driven approach to continuously reconcile the pressures of demand and supply of workspace, there must be a channel for continuous dialogue between the strategic management of core business development, and their colleagues responsible for operational management of business resources. It is clearly desirable to have an integrated management framework that incorporates considerations of facilities provision and their ongoing management as an integral part of the business planning cycle. Figure 5.5 illustrates the use of the *Strategic Facilities Brief* (SFB) and the *Service Levels Brief* (SLB) as instruments for promoting and maintaining the critical interface between strategic management decisions and operational decisions.

The strategic facilities brief (SFB) is the output document that defines the operational needs emanating from the organisation's business plans. The principal purpose of the SFB is to define a corporate procedure which guides key facilities attributes and service performance criteria that are required to fulfil the organisation's objectives as dictated by the business plans. The source of the strategic inputs is *business information*. The service level brief (SLB) represents the definition of acceptable performance levels in respect of the physical asset base and the requirements for support services as defined by the SFB. The principal purpose of the SLB is to define and quantify the appropriate support services and facilities infrastructure supporting the activities of the business units, and their performance within the workplace environment. The role of the SLB is to ensure that the appropriate facilities and support services are delivered at the appropriate operational level as defined by the SFB. The source of the tactical inputs is *facilities information* (Then, 2003a).

Figure 5.5 Reconciling supply and demand to create an enabling workplace environment (Then, 2003b).

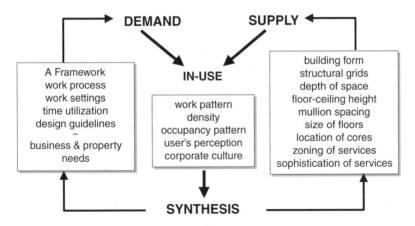

Figure 5.6   An objective space analysis methodology (adapted from McLennan, 1993).

Underpinning any development of a facilities and real estate strategy is the requirement for a robust methodology based on objective analysis. In this respect, the conventional 'supply-led' approach that focuses on building and estate management aspects must be preceded by a realistic quantification of the business's demand drivers. In other words, the matching process must be driven from *demand to supply*, not from *supply and fit*. It is part of a structured approach to reconciling the demand and supply factors, defined in both quantitative and qualitative terms, and ultimately reflected in the quality and effectiveness of the workplace environment and its layout.

An objective demand-led approach starts with the foundation of the *space budget* derived from objective organisational analysis of space demands. The process should start by determining the key requirements of the organisation which should be translated into the desired building attributes of shell, core, shape, infrastructure and services. The progression from *space budget* through the intervening levels of *planning concepts*, *furniture and equipment*, *building performance* and *image and design*, facilitates a logical consideration of issues that must be assessed before a supply decision is taken on any individual building, existing or new. Figure 5.6 illustrates the process of reconciliation of demand and supply via a methodology that evaluates demand from a framework derived from business and property needs. It indicates the role of building supply specifications, current practices and the perceptions of workspace in use. The proposed methodology provides a structured approach which takes into consideration the supply and demand factors described above.

5.7   MAINTAINING STRATEGIC RELEVANCE

In terms of space planning and management, maintaining strategic relevance is about establishing a baseline position wherein the gap between demand for and supply of space, is kept to a minimum. For the workplace to function as a strategic element in the enterprise, it must be recognised that its success depends on the dynamic interrelationship and mutual reinforcement of spatial, organisational, financial and technological issues (McMorrow, 1999). It is important to achieve an active coherence between the required work and these aspects of the workplace — they must be aligned. Alignment ensures workplace is in context, that it supports and reflects an organisation's business strategies, economic realities and organisational culture, as well as demographic and social changes.

5.8   THE NEED FOR DIALOGUE

It is clear from Figure 5.5 above, that the development of the SFB and the SLB involves participation and inputs from the strategic business planning process, together with contributions from real estate/facilities management and the participation of users of workspaces. The interactions between the SFB and SLB are necessary to ensure proper matching of demand for facilities, with the supply of appropriate facilities and services necessary to support the implementation of the desired business plans. Proactive management of operational assets requires clear strategic direction from the business's senior management and clear measurable deliverables from the operational management team.

The process of matching supply to demand is clearly never easy, and is frequently a complex one. The final outcome is an appropriate structure of operational assets that is aligned with current corporate objectives, while at the same time acknowledging that at any point in time, the goal is to optimise the use of facilities and services such that a balance is achieved between demand and supply — the *current steady state*. The concept of the current steady state is central to the effective management of operational assets and their associated services over time. The emphasis on '*adjustments to the next steady state*' recognises the volatile environment of the business world in which facilities managers must operate as proactive agents for change (Then, 2003).

5.9   MANAGING OCCUPANCY COST — MONITORING UTILISATION

All businesses require real estate or facilities to operate from. Alexander (1998) states that facilities are commonly an organisation's second largest expense and can account for up to 15% of turnover, and typically constitute the largest item on the balance sheet. Yet the fact that it is one of most undermanaged corporate assets has not changed since the early 1980s when the issue was highlighted by two studies from Harvard University (Zeckhauser and Silverman, 1981) and MIT (Veale, 1989):

> ... corporate real estate, the buildings and land owned by companies that are not primarily in the real estate business, typically accounts for 25 percent or more of a firm's total asset. Despite this enormous value, it remains an under-managed asset. (Zeckhauser and Silverman, 1981)

> One of the most significant conclusions of the MIT study is that many corporate real estate managers do not have adequate information on their real estate assets. The lack of informed decision making and awareness characterized nearly every dimension of corporate real estate management examined. In general, under-management is a better descriptor of the situation than mismanagement. These assets are not necessarily managed poorly but are, in many cases, not managed to their full potential. (Veale, 1989)

In a recent article in the Harvard Business Review (Nov. 2009), it was quoted that:

> For most businesses, it [real estate] is the largest or second-largest asset on the books, yet, because it is everywhere, real estate is easy to take for granted. ... Business real estate is not merely an operating necessity; it's a strategic resource. But it rarely captures senior management's attention. (Apgar, 2009)

Even after almost three decades, there appears to have been little change in that corporate real estate assets continue to be under-managed. As a result, sub-optimisation in the utilisation of the real estate resource still appears to be prevalent in many organisations. However, the pressure to reduce operating costs has, in recent years, shifted from the initial focus on reducing overall staff numbers (headcount reduction) to initiatives aimed at reducing premises occupancy costs. The pressure to reduce occupancy costs has resulted in the scrutiny of two main aspects — the amount of space occupied by business units and the utilisation of existing spaces.

Utilisation is about achieving the optimum match between timing of need and availability of supply. The objective is to aim for the best match between demand for workspace to support business activities and its availability in terms of timing and duration of requirements. Demand for workspace relates to appropriateness of the types of space and their distribution within a building or a portfolio of buildings. Supply relates to the pattern of use of the available space and the range of tasks it is supporting.

No matter how well a workplace is engineered to meet the needs of its users, it adds no value to the business when not in use. Experience from many office workplace utilisation surveys shows that, for activities which are 'desk bound' occupants rarely spend more than 80% of their time actually at their workplace, i.e. a desk. In practice, it is not uncommon to encounter many businesses where the utilisation of their workplaces falls between 40% and 50%. Support for this statistics is substantiated by a recent interview with Philip Ross, CEO of Cordless Group.

> If you take an average snapshot of an office across the world and across industries, about 50—55 per cent of desks in offices are empty at any one time but at the same time you can't find a free meeting room. ... there is a complete mismatch between the space provided today and what people do in an office — there is much greater demand for team-working spaces, where individuals work together on projects and pitches and a far declining demand for people to sit behind the same desk day-in day-out. (FM World Daily, 14 April 2010)

Such situations have led more and more businesses to adopt 'flexible' styles of working or 'alternative workplace', where the focus is on providing a variety of work settings in which the user is free to choose (Apgar, 1998).

Utilisation is not only a function of efficiency of layout, but is also derived from the degree of flexibility the premises afford the users in terms of its capability to support the relocation of working groups, departments and functions (i.e. churn), and also to meet changing patterns and practices of work. Provision for flexibility can be made by consciously recognising that workspace demand at the portfolio level can decrease as well as increase. Through workspace design it is possible to incorporate features that afford variety in meeting the needs of customers.

MANAGING SPACE DEMAND OVER TIME

Future role of work and workplace design

In recent years, there have been several speculative studies relating the future of work (Ratchiffe and Saurin, 2007, 2008; CMI, 2008a,b; Future Foundation, 2010; IBM Institute for Business Value, 2010). The research approaches adopted for these reports were similar in that the primary objective was to foster and promote a more informed, structured and imaginative understanding of the long-term strategic stewardship of the built environment. This was achieved through visioning scenarios of what the future of work and management looks like and examining how business can prepare for it. In short, the reports aimed to identify the forces that will change and influence the world of work in the future.

The central themes of these reports acknowledge that:

> ... workplace change and innovation has become critical to the future of organisations in a dynamic, economy-driven and knowledge-based society. Managing this change, however, is a vital dimension underpinning successful transition — to new work styles, patterns and locations all within the aegis facilities management. Providers of physical and virtual workspaces need a clear understanding of the forces driving these changes and their impact, not only on individuals, but also on the organisations themselves. (Ratchiffe and Saurin, 2007)

The Executive Summary of the CMI (2008b) Report on *Environmental Scanning — Trends affecting the world of work in 2018,* provides us with glimpses of the future of work grouped under three groups of possible trends:

Trends influencing the world of work:
- increasing pressure for global competition
- growing opportunities of a global marketplace
- an increasingly connected and accountable world.

Trends influencing the structure of work:
- workplace regulation that promotes and protects equality
- the rise of flat organisations
- increasing pressure to train
- the rise of multi-generational workforce.

Trends influencing the way people work:
- varied career paths
- changing employment contract
- changing expectations of work
- interpersonal skills
- changing working hours
- new technologies.

It is clear from the above review that changes in the workforce and the nature of work are inevitable. The key variable is how businesses will adapt to the changes and the pace of change they choose to adopt. For real estate and facilities management professionals, one of the main challenges is to develop the competencies that are necessary to facilitate the business cases for transforming current practices and for managing the changes to their workplace portfolio and facilities services that will result in improved overall corporate performance.

5.10.2   Implications on workplace management

According to a recent study by the IBM Institute for Business Value (2010), organisations that are significantly outperforming their industry peers also happen to be making more headway on newer approaches to work. The most dynamic, collaborative and connected companies have widely adopted specific technologies that make smarter working practices viable. In short, it pays to work smarter not harder. Based on input from executives from around the world, the same study also revealed that the motivation behind smarter working practices is not just based on efficiency, but also on long-term growth — growth fuelled by agility, i.e. having the capability to generate innovative ideas, spot opportunities and then act on them.

Brennan (2010) speculates that the current economic downturn has created 'the perfect storm for a mobile workspace to really gain traction.' He concludes that '. . . the evolution of the workplace, driven by technology and mobility, will propel a paradigm shift in the way in which we evaluate, acquire, utilize and dispose of real estate.'

In the context of workplace, agility means more than just having buildings and communication technology ready for alternative uses. It means continuously improving work and the infrastructure that enables it. Agility is defined here as 'the ability to respond quickly and effectively to rapid change and high uncertainty.' That agility is achieved 'through the co-evaluation of the workplace and work' (Gartner and MIT, 2001) — in my opinion, the very core of facility space planning and management.

From the above review, it can be observed that more recently, facility space planning and management emphasis has shifted from provision to utilisation, and from alignment to agility and flexibility. Facility professionals must 'think in terms of space demand probabilities, and be able to reconfigure workplaces to reflect changing workforce needs for space on almost a daily or even hourly basis.' (Ware and Grantham, 2010).

Technology, in its broadest sense, will continue to be a major driver of business and social change (Ernst & Young, 2010) but its impact on the nature of work will be even more profound than it is today. Changing demand variables (such as the impact of remote and distributed working; multi-generational workforce, work-life balance, corporate social responsibilities and environmental sustainability, etc.) — present pressures that will continue to confront the real estate/facilities professionals. Rather than viewing the future as a threat, professionals in the field of real estate and facilities management should seize the impending change as an opportunity for enhancing their credentials as effective change managers of the corporate enabling environment. A recent article describing Cisco's next generation workplace appropriately expressed this sentiment in its subtitle: 'Don't simply anticipate change, *drive* it.' (Stritch and Sept, 2009).

In management, the choice is never between planning and not planning. What is important is the kind of planning we do. Successful change management applied to workplace management is about managing the dynamics of change from all dimensions — economic, technological, political and social. A piecemeal approach to planning change in workplace has been disappointing.

The key is in the point of view, a change in mindset or framework, not the details. When we make the leap, we see everything differently and can operate in a powerful system for holistic planning and managing long-range, creative and constructive change. (Atkinson, 2008)

The business of managing space in corporate facilities needs to embrace this change in mindset.

5.11   ACKNOWLEDGEMENTS

The content of this chapter is a personal synthesis of ideas and concepts from numerous published sources, books, journal articles and websites, that are listed below in the References, to which the author owes his acknowledgement. However, special mention of acknowledgement must be attributed to Wes McGregor, the co-author of our book *Facilities Management and the Business of Space*, in which Section 5.2 on *The Space Planning Process* represents a synthesis from our earlier work.

REFERENCES

Acoba, F.J. and Foster, S.P. (2003). Aligning corporate real estate with evolving corporate missions: Process-based management models. *Journal of Corporate Real Estate*, 5(2), 143–164.

Alexander, K. (1998). 'Facilities management: A strategic framework'. In *Facilities Management: Theory and Practice*, ed. K. Alexander, E. & F. N. Spon, pp. 2–11.

Apgar IV, M. (1998). The alternative workplace: Changing where and how people work. *Harvard Business Review*, 76(3), 121–136.

Apgar IV, M. (2009). What every leader should know about real estate. *Harvard Business Review*, November, 100.

Apgar IV, M. and Bell, M. (1995). Managing real estate to leverage change: The Dun & Bradstreet case. *Site Selection*, December, 935–938.

Arthur Anderson (1995). *Wasted Assets? A survey of corporate real estate in Europe*.

Atkinson, A. (2008). Making the quantum leap through strategic results. *The Leader*, May/June, 82.

Avis, M. and Gibson, V. (1995). *Real Estate Resource Management*. GTI.

Becker, F. and Steele, F. (1994). *Workplace by Design — Mapping the High-performance Workscape*. Jossey-Bass, San Francisco.

Brennan, R.J. (2010). A paradigm in transition: The next generation of real estate and workplace. *The Leader*, March/April, 12–18.

Cairncross, F. (2001). *The Death of Distance*. Harvard Business School Press, Boston.

Chartered Management Institute (CMI), UK (2008a). *Management Futures — The World in 2018*.

Chartered Management Institute (CMI), UK (2008b). *Environmental Scanning — Trends Affecting the World of Work in 2018: Executive Summary*, pp. 12–15.

Ernst & Young (2010). *Business Redefined — A Look at the Global Trends That Are Changing the World of Business*. Ernst & Young Global Limited.

Froggatt, C.C. (2001). *Work Naked*. Jossey-Bass, San Francisco.

Future Foundation (2010). *Visions of Britain 2020 — The Workforce*.

Gallup Organisation (1996). *Shaping the Workplace for Profit*. Commissioned by Workplace Management.

Gartner and MIT (2001). *The Agile Workplace: Supporting People and Their Work*, pp. 20.

Graham Bannock & Partners Ltd. (1994). *Property in the Board Room - A New Perspective*. Commissioned by Hillier Parker.

Grantham, C. (2000). *The Future of Work*. McGraw-Hill, New York.

Haynes, B.P. and Nunnington, N. (2010). *Corporate Real Estate Asset Management: Strategy and Implementation*. Elsevier.

IBM Institute for Business Value (2010). *A New Way of Working — Insights From Global Leaders*.

Joroff, M., Louargand, M., Lambert, S. and Becker, F. (1993). *Strategic Management of the Fifth Resource: Corporate Real Estate, Report of Phase One: Corporate Real Estate 2000.* The Industrial Development Research Foundation, USA, p. 7.

Malone, T.W. (2004). *The Future of Work.* Harvard Business School Press, Boston.

McCann, M. and Venable, T. (2010). Microsoft's Chris Owens: Leading real estate in a fast-paced environment. *The Leader,* January/February, pp. 42.

McGregor. W. and Then, D.S.S. (2001). *Facilities Management and the Business of Space.* Butterworth-Heinemann, London.

McLennan, P. (1993). *A Facility Planning Framework for Programming the Office Environment,* In *Professional Practice in Facility Programming,* ed. Preiser, W. Van Norstrand Reinhold, New York, pp. 350–351.

McMorrow, E. (1999). CIR, ERP are in your future. *Facilities Design and Management,* 18(6), 9.

Nourse, H.O. and Roulac, S.E. (1993). Linking real estate decisions to corporate strategy. *The Journal of Real Estate Research,* 8(4), 475–494.

Olesand, N. (2008). The evolving workspace. *PFM,* October, 14–16.

Osgood, R.T. (2009). *Key Issues in Enterprise Alignment: Survey Results and Practical Experience with the Fortune 1000.* Flad Architects and CoreNet Global.

Perske, K., Jordon, R., Gillsepie, C. and Sanquist, N.J. (2009). *Mobility: The New Workplace Imperative – A Call to Action for Business, People and the Environment.* A position paper by Group 5 Consulting and Manhattan Software, pp. 4.

Prasow, S. and Sargent, K. (2009). Planning for uncertainty: Strategies for today and beyond. *The Leader,* Nov/Dec, 12–16.

Ratchiffe, J. and Saurin, R. (2007). *Workplace Futures – A Perspective Through Scenarios.* Johnson Controls Facilities Innovation.

Ratchiffe, J. and Saurin, R. (2008). *Towards Tomorrow's Sustainable Workplace – Imagineering a Sustainable Workplace Future.* Johnson Controls Facilities Innovation.

Ross, P. (2010). A brave new world. *FM World Daily,* 14 April. http://www.fm-world.co.uk/features/interviews/fm-interview-philip-ross//locale=en Accessed 3 Aug 2010.

Stritch, P. and Sept, C. (2009). Cisco's next generation workplace: Don't simply anticipate change, drive it. *The Leader,* May/June, 22–25.

The Henley Centre (1996). *The Milliken Report: Space Futures.* Commissioned by Milliken Carpet.

Then, D.S.S. (1994). Facilities management – The relationship between business and property. *Proc., EuroFM/IFMA Conference on Facility Management European Opportunities.* Brussels, Belgium, 253–262.

Then, D.S. S. (1999) An integrated resource management view of facilities management. *Facilities,* 17(12/13), 462–469.

Then, D.S.S. (2003a). Strategic Management. In *Workplace Strategies and Facility Management – Building in Value.* Ed. R. Best, C. Langstonand G. de Valence. Butterworth-Heinemann, UK, pp. 69–80.

Then, D.S.S. (2003b). Integrated resources management structure for facilities provision and management. *ASCE Journal of Performance of Constructed Facilities,* 17(1), 34–42.

Then, D.S.S. (2005). A proactive property management model that integrates real estate provision and facilities services management. *International Journal of Strategic Property Management,* 9, 33–42.

Veale, P.R. (1989). Managing corporate real estate assets: current executive attitudes and prospects for an emergent management discipline. *Journal of Real Estate Research,* 4(3), 16.

Ware, J., Grantham, C. (2010). 21st Century Space Planning. *Future of Workplace Agenda,* (http://www.thefutureofwork.net/assets/21st_Century_Space_Planning_July_2010.pdf) Accessed 20 August 2010.

Wilson, R.C. (1991). Strategic positioning and facilities planning: reviewing business plans and facilities strategies. *Site Selection. Industrial Development Section,* April, 26–30.

Zeckhauser, S. and Silverman, R. (1981). Corporate real estate asset management in the United States. Harvard Real Estate, Inc. pp. iv.

6 Project Inception: Facilities Change Management in Practice

Jim Smith and Peter Love

CHAPTER OVERVIEW

The production of buildings moves through a number of stages involving a large number of participants. At the early stages of a project key strategic decisions are made. The decisions made during the *strategic* or early design stages in the life of a project are seen as a critical factor in influencing the fundamental characteristics of quality, cost and time of projects. Research in many countries has identified the need for clients and their advisers to be aware of the importance of what can be commonly termed the project inception stage, where the *strategic level of decision-making* is focused. The authors consider that facilities managers have the expertise and capability to make a valuable contribution to the project inception stage. Blending their basic skills with a broader appreciation of the client organisation's direction can provide a value-adding service to an expanding range of clients. The authors believe that a range of critical factors needs to be taken account of in the design of a rigorous and successful project inception process and these can be summarised as:

- recognition of the true nature of the organisation's business or facilities problem and that its environment is complex and fluid
- strategic management needs to be integrated into the project inception process
- stakeholder participation should be encouraged throughout the process, as it is essential to gain their views and opinions on the type and quality of the outcome they are committed to, whether it results in a building or a non-building solution
- the decision to build and strategic management should be integrated to ensure compatibility of the built facility with the strategic direction of the organisation expressed in the business case.

Keywords: Clients; Project inception; Strategic decision-making.

6.1 INTRODUCTION

In the previous chapter Then shows that the rationale for decision making regarding space planning in a business needs to be multi-dimensional, if it is to be successful.

Facilities Change Management. Edited by Edward Finch.
© 2012 Blackwell Publishing Ltd. Published 2012 by Blackwell Publishing Ltd.

Space-planning decisions must be linked and integrated with the business case and take account of technological factors supporting flexibility and mobility, corporate image and culture. The focus must be on involvement of the users and a management capability that has a long-term view of the organisational needs. Whilst this is a demanding process, a dynamic organisation must embrace the challenges of this more volatile business environment.

The production of buildings moves through a number of stages involving a large number of participants. The decisions made during the *strategic* or early design stages in the life of a project are seen as a critical factor in influencing the fundamental characteristics of quality, cost and time of projects (Walker, 1995). The contributions from the formally constituted client, design and construction teams charged with the responsibility of delivering the project are crucial to the success of a project. However, it is increasingly recognised that external participants such as users, customers and members of the community may have a useful role to play in influencing the location, form, content and timing of the project. In fact, it is now appreciated that the multiple attributes that contribute to the success of a project (financial, social, technical, functional, aesthetic, environmental, political) are influenced by a whole host of decisions by various individuals, bodies and organisations (Property Services Agency, 1981). It can be categorised as a complex process (Walker, 2002).

More research in the UK has identified the need for clients and their advisers to be aware of the importance of what can be commonly termed, the *strategic level of decision making* (Keel and Douglas, 1994). Atkin and Flanagan (1995) conducted a survey of construction clients that indicated that the strategic level had the most potential for cost savings in a project. The importance of the strategic stages in the development of solutions was further reinforced by the Royal Institution of Chartered Surveyors QS Think Tank (CSM, 1998a).

The Think Tank (CSM, 1998a) stated that client advisers should place high priority on:

- understanding the project priorities and business objectives
- providing advice which assists clients to gain competitive advantage and
- being client orientated rather than being focused too much on the details of the project to the detriment of the broader issues and objectives.

They also considered that a growth area in the future would be project strategy work, although they conceded that few design professionals have yet established a strong position in this area (CSM, 1998b). Rather than 'design professionals', the authors suggest that facilities managers have the skills, knowledge and capability to carry out these important activities in the formative stages of a project. Facilities managers have a broad view of the problem and the potential solutions, some requiring a project and others not.

Whilst many clients focus on cost savings, such savings are only a part of the complex project equation. The less tangible factors such as aesthetics, complementary interacting functions and positive unintended consequences of the development that contribute to the quality, value and effectiveness of a project are equally, if not more, important. The potential of these early stages, particularly the strategic stage of inception, is substantial when compared to decision making later in the process, when the effect of change is only marginal.

So, in theory, practitioners should seek to devote more time and apply more rigour to the decision making during the formative stages of a project, particularly the project inception stage. However, in practice, some obstacles have to be faced. The culture amongst some design professionals and amongst many clients often creates a climate of expediency. The consequence of this is a tendency to rush the early stages of projects and to under-exploit opportunities in the initiation or inception stage. Unfortunately, this short-term thinking does not always lead to a soundly based strategy to guide a well thought out project and facilities solution.

In order to improve the process of project inception, the client project team needs to look more closely at the nature of the problem on which we have decided to focus. Rather than a rigid framework, the client and the project team requires a more general, flexible schema to provide the framework for this approach. Using a problem-solving approach to guide our work may be more useful. This can provide us with guidance and insight into traps lurking within the organisational and design environment. In general terms, a problem is a question in need of a solution. In the context of facilities management, someone, or a group, has identified an unsatisfactory state of affairs, which has created a need. This need may be solved by the construction of a new building, the extension of an existing building or a re-arrangement of the activities within an existing building, or some combination of these (Mohsini, 1996).

6.2 PROJECT INCEPTION

One of the crucial components of the project-briefing stage is the first step in the development process, now termed *project inception*. An alternative term sometimes used is *project initiation*, but the term 'project inception' is used throughout this chapter. The design and construction disciplines have long recognised that the early stages in the development of a project are crucial to its success. The reason is that the most significant decisions are made at the inception or pre-design stage and such decisions will influence the characteristics and the form of the project (Allinson, 1997; RIBA, 2000; Kirkham, 2007). Once these significant decisions have been made, by their very nature, they cannot be readily reneged or dramatically altered in the subsequent stages. Therefore, if these early stages are so crucial to the success of a project it is important that these decisions are properly considered. Ideally, these stages should also attract sufficient time, resources and expertise to be carried out conscientiously.

The types of decisions made during the early stages have a dramatic effect on the more tactical aspects of a project: *cost, time* and *quality* objectives. These are discussed and analysed at length in the building economics literature by Morton and Jaggar (1995), Seeley (1996), Kirkham (2007), Smith and Jaggar (2007) and Ashworth (2008).

Research in the UK has also identified the need for clients and their advisers to be aware of the importance of what can be commonly termed the *strategic level of decision making* (Keel and Douglas, 1994; Latham, 1994). Atkin and Flanagan (1995) conducted a survey of construction clients that indicated the strategic level (client's business case) had the most potential for cost savings in a project, at 10–20%. The tactical (5–15%) and the operational levels (1–10%) showed fewer opportunities for savings.

It seems reasonable to assume, based on experience, that better results will be developed when we broaden the source of ideas for solutions. The possibility of a *breakthrough*, innovative or creative idea or just a better approach to facilities provision is more likely to occur if we create a more diverse, broad or mixed environment at the project inception stage. So, the client and other stakeholders need to play a key role. Whilst the client is a significant player in guiding the evolution of alternative strategies, it must be emphasised that all participants should be tolerant towards alternative solutions (Woodhead, 2000; Karma and Anumba, 2001; Woodhead and Smith, 2002). Some strategies and solutions may not match their preconceptions about the expected future direction of the project(s) they have in mind. Their 'pet' projects or 'comfortable' solutions may be questioned. The process must also be useful, flexible, well organised and sensitive to client and stakeholder objectives.

The *project inception* stage, which straddles the strategic and tactical levels, was largely ignored or neglected in the professionally based models (AIA, 1987; RAIA, 1993; RIBA, 1998). The exception is the work of AIPM/CIDA (1995). An obvious reason for this neglect is that such models of the development process aim to guide design and construction team activities. Traditionally, project inception involves insufficient design team members to warrant detailed attention. In addition, this early stage may be so amorphous and indefinable that these professional bodies cannot be precise enough to provide a useful guide for users and practitioners. Some would also suggest that since this stage is often speculative and ill defined, the professional bodies appear to have little regard for defining and regulating the possible activities. Therefore, each project will have unique circumstances that defy such a prescriptive approach favoured by the professional bodies. Eventually, the RIBA (2000) has revised its *Plan of Work* to recognise the inception or feasibility stage by including a new stage, *Establishing the Need*. This Plan of Work is summarised in Table 6.1. This pre-design stage is focused on the client, the project and a building, but the activities lack the essence of looking at the client's problem in its totality before deciding a building is required. This is where facilities managers can make a more important contribution. Their global view of the client's problem allied with their knowledge of facilities, functions and performance enables a more holistic solution.

6.3   DEFINITION OF PROJECT INCEPTION

Project inception has been given other titles. For instance, Atkin and Flanagan (1995) have termed it the *strategic stage* or the *client's business case*. Latham (1994) called it *project strategy* and *need assessment*. Others such as Palmer (1981) and RAIA (1993) call it *pre-design*. White (1991a,b) terms it *facility programming* and Mohsini (1996) has the more poetic term *front-end incubation*. CIDA (1993), AIPM/CIDA (1995) and Best and Valence (1999) have used the term *project initiation*. As noted earlier, we have decided to use the term *project inception*. As noted above, the RIBA (2000) term the stage 'Establishing the Need'. However, irrespective of the title used, all agree that it is the stage at which the *decision to build* (Woodhead, 2000) or *commitment to proceed* (AIPM/CIDA, 1995: 18) has been made and the strategic direction of the facilities and the project will be decided.

International research has focussed on the project initiation stage as the most critical part in a project's life cycle (CIDA, 1993; Latham, 1994; AIPM/CIDA, 1995; DIST, 1999). In recent times, in the UK and Australia, clients have expressed dissatisfaction with the process and the products of the construction industry (BPF, 1983; NPWC/NBCC, 1990; Latham, 1994). An important feature of the aforementioned work in these countries is that the early stages of a project have received special attention.

In Australia, the work of CIDA (1993) made it one of the first bodies to define the project inception stage. This was further refined in a joint approach with the Australian Institute of Project Management (AIPM/CIDA, 1995) to produce a reformulated definition with greater detail of the phases, activities and outcomes during the project initiation (inception) stage. Figure 6.1 summarises the activities and decisions made at each of six phases during the project initiation stage as envisaged by AIPM/CIDA (1995).

The importance of this work is that it gives due prominence to the strategic aspects in the development of a project in the planning and concept phases (project inception). Thus, the work by AIPM/CIDA (1995) and Atkin and Flanagan (1995) appear to parallel each other in concept. This AIPM/CIDA model of the process is a useful reference point for the

Table 6.1 RIBA pre-tender work stages in outline Plan of Work. The RIBA Outline Plan of Work 2007 (© Royal Institute of British Architects) is reproduced here with permission of the RIBA.

| | PRE-DESIGN | | | DESIGN | | |
| | Inception or Feasibility | | | Pre-Construction period | | |
Pre-Stage A	**Work Stage A** 1	**Work Stage B** 2	**Work Stage C** 3	**Work Stage D** 4	**Work Stage E** 5	**Work Stage F** 6
Establishing the Need	Options Appraisal	Strategic Briefing	Outline Proposals	Detailed Proposals	Final Proposals	Production Information
Activities in each Stage						
Client has to establish client management team, appoint client representative, appoint cost consultant. Identify objectives, physical scope of project, standard of quality of building(s) and services, time frame and establish budget	Identify client's requirements. Possible constraints on development. Prepare cost and VM studies to enable client to decide whether to proceed. Cost of preferred solution. Select the probable procurement method. Instruct development of preferred solution to strategic brief stage. Prepare outline business case.	Prepare strategic brief. Confirm key requirements and constraints. Identify procedures, organisational structure and range of consultants and others to be engaged for the project. Target cost and cash flows. Whole life costing.	Evaluate strategic brief with a consideration of time, cost, risks and environmental issues. Establish design management procedures. Prepare initial cost plan, project programme, cash flow. Develop project brief. Estimates of cost prepared.	Evaluate outline proposals, complete and agree user studies. Complete and sign off on project brief. Receive design and cost input from team and develop detailed design solution. Firm cost plan and cash flow projection. Development control submission. Review procurement advice.	Sanction and complete final layouts. Coordinate all components and elements of the project. Cost checks design against cost plan. Decide on procurement method. Review design and cost plan. Prepare submission for statutory approvals. Design is now frozen.	Prepare all coordinated production information including location, assembly and component drawings, schedules and specifications. Applications for statutory approvals completed. Provide all information for final cost checks of design against cost plan.

Source: RIBA (2000).

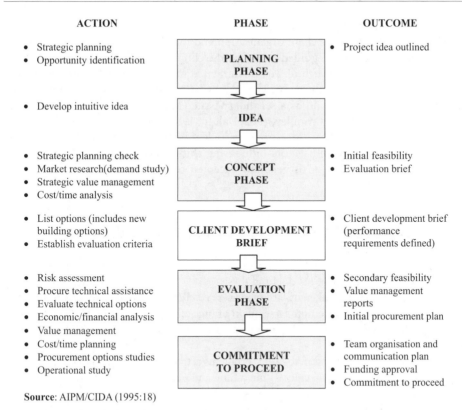

ACTION	PHASE	OUTCOME
• Strategic planning • Opportunity identification	**PLANNING PHASE**	• Project idea outlined
• Develop intuitive idea	**IDEA**	
• Strategic planning check • Market research(demand study) • Strategic value management • Cost/time analysis	**CONCEPT PHASE**	• Initial feasibility • Evaluation brief
• List options (includes new building options) • Establish evaluation criteria	**CLIENT DEVELOPMENT BRIEF**	• Client development brief (performance requirements defined)
• Risk assessment • Procure technical assistance • Evaluate technical options • Economic/financial analysis • Value management • Cost/time planning • Procurement options studies • Operational study	**EVALUATION PHASE** **COMMITMENT TO PROCEED**	• Secondary feasibility • Value management reports • Initial procurement plan • Team organisation and communication plan • Funding approval • Commitment to proceed

Source: AIPM/CIDA (1995:18)

Figure 6.1 Building project life cycle: project initiation phases and stages. Reproduced here with permission of the Australian Institute of Project Management (AIPM).

development of any framework for this stage. However, inspection of the first two stages (planning and idea) indicates a lack of detail of activities (action) and output (outcome). These are the areas where techniques and approaches are needed that assist in developing the initial statement of client requirements contained in a performance brief (it may also be known as 'evaluation brief', 'needs statement' and 'client development brief' in other systems) to guide the concept phase and later stages. However, of significant importance is that these initial stages should attempt to provide a link between strategic management of the organisation and this project inception.

6.4   THE DECISION TO BUILD

After taking account of all the strategic factors and the organisational environment, the decision to build represents a strategic initiative or assessment to embark on a built solution to satisfy the organisation's strategic objectives. It is a decision that members of the design team should be aware of, reflect on and consider why the decision to build was made. In this way, their advice and activities during the design stage should be more relevant to client needs. By understanding the history of the decision, the design team may be better able to respond to client requirements and opportunities (Bilello, 1993; Woodhead, 1999, 2000).

However, most participants in the design process are often divorced from the decision to build within the client organisation and may lack the necessary background information that informed, guided or forced the choice of a building solution (Bilello, 1993). In fact, there is a dearth of well-documented material and literature on this key decision in the development process. The literature that does exist often ignores or only includes the decision to build as a secondary issue (Ministry of Public Buildings and Works, 1970; Weeks, 1980; Bromiley, 1982; Dalton, 1989; Chiu, 1991). Few authors have documented this part of the process and participants in the process seem remarkably indifferent to the forces and factors influencing this crucial decision. The attitude is often one where the decision is generally accepted as a *given*, not to be revisited or amended by later players.

6.5 FRAMEWORK FOR THE DECISION TO BUILD

Woodhead (1999) has conducted the most thorough and comprehensive analysis of the decision to build. The aim of his doctoral thesis was, '. . . to understand the influences on the decision to build undertaken by large experienced clients of the construction industry in the UK . . .'. It specifically attempted to explain a range of issues in relation to the decision to build: the process, the content, what influences it, why those influences affect the process and content.

The essence of client 'decision to build' problems is that they are often not obvious, but a solution must be found and usually within a time and possibly a cost constraint. Often the goal is unclear or not properly stated. This can lead the designer into a fruitless search up and down the hierarchy of the problem by means of *escalation* or *regression* (Lawson, 1995). This may be frustrating to the client and the participants, but it may lead to an *insightful* solution (Rickards, 1990) that dissolves the problem (Ackoff, 1978).

Thus, the objective of the project inception stage must be to search for the real cause of the problem and not to merely accept a solution that is a prescribed response which fails to uncover the obvious and latent issues involved. In many cases, and increasingly in this technological age, the solution to the client's problem may not always be a building or buildings.

6.6 GAPS/DISCONTINUITY IN THE PROCESS

The literature (NPWC/NBC, 1990; Latham, 1994; Egan, 1998; Smith and Jaggar, 2007) has noted an established gap between the design process and the construction activities further down the development process (or the supply chain as the more recent literature terms it). However, this research is now pointing out a second gap between the strategic management in an organisation and the design development activities that follow. That is, through our case studies it was observed that an equally serious discontinuity inhibits the attainment of the best solution and performance for a proposed project. A number of authors (Drage, 1970; DHSS, 1986; Hughes and Walker, 1988; Hughes, 1989; Walker, 2002) have indicated that there is a gap between the pre-design activities and the design stages. This research confirms that such a gap exists in a tangible form.

In contemporary terms the supply chain exhibits discontinuity at two critical points: firstly, when a project passes from the strategic 'decision to build' to the design stage and secondly, when it is translated from a design *model* into its built form at the construction stage. The consequences of this gap (after the decision to build has been taken at the strategic level) are that it is not properly or fully communicated to the design team. The design team naturally fills any information voids and makes its own assumptions to provide the necessary performance and design characteristics. So, when the decision passes to the design team it may result in a project decision that occurs in a vacuum. It may be convenient for the design team to begin the process with a *clean sheet*.

Making progress to closure of the gap (or reduction) between the strategic decision-making and design stages needs the cooperation of the senior management that makes and implements the decision to build. It is essential that strategic factors and stakeholder participation are requisite characteristics of the process. That is, in the process of transfer from the strategic management context to the 'decision to build' and project inception, the senior management should not abdicate its responsibilities and but should remain in close contact with the next critical stage.

A proposed solution to provide greater integration of the strategic activities with the project inception activities is shown in concept form in Figure 6.2. This figure brings together the disparate elements of decision making within an organisation at the various levels and attempts to integrate it into a framework that recognises that the 'decision to build' is part of the strategic and operational environment. It identifies a range of possible approaches to assist in this process of integration and these are shown on the right-hand side of the diagram.

Approaches such as that adopted by strategic needs analysis (SNA), which has been used by the authors, may assist in the process of bridging the gap between the strategic objectives of the organisation with the 'decision to build' by encouraging management commitment to the process and the decision. This technique has been described by one of the authors in a number of publications. (Smith *et al.*, 2003; Smith and Love, 2004, Smith, 2005a,b). However, unless there is senior management support for the process then their decision cannot be guaranteed. No process, technique or approach can ensure management commitment, especially when the eventual decision from SNA (or any other project inception technique) is not favoured, or it reduces the capability for flexible decision making.

6.7 MODEL OF THE PROJECT INCEPTION PROCESS

After the decision to build, or not to build, has been made then a project inception process is necessary that converts the strategic management decision into practice. A model of this process is proposed which builds upon the structure first presented in Figure 6.2, and is shown in Figure 6.3. This is a model of the project inception processes. It is a summary of the activities and related activities in: (1) creating the options; and (2) making a choice of the strategic direction that the project will take as a result of stakeholder involvement and decision making. Naturally, after its development and testing by the authors, they believe that techniques such as SNA can deliver a better way of defining client and stakeholder requirements. These can be captured in a performance brief that properly describes the client group needs and requirements in a form that gives direction to the design team, but does not hinder their creativity nor their ability to explore alternative ways to satisfy the strategic performance requirements documented more explicitly by SNA (Smith and Love, 2004; Smith, 2005a,b).

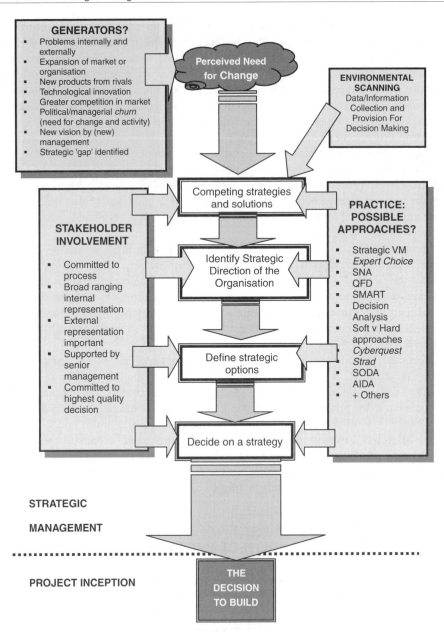

Figure 6.2 Model of the project inception process: strategic.

6.8 PERFORMANCE BRIEFING

The conversion of these characteristics into a working document for use by clients (stakeholders) and the business and design team needs the use of a strategic document that can bridge the gap between these groups. This strategic brief, with an emphasis on the required performance of the project, needs to use a format, style and language all

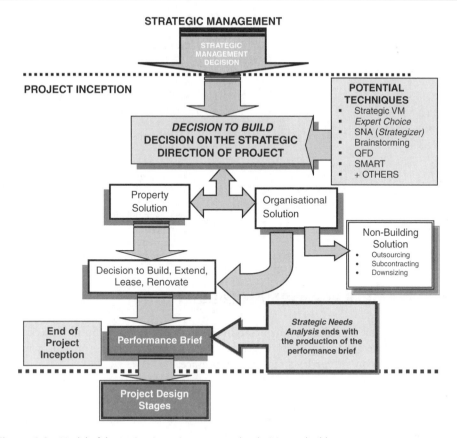

STRATEGIC MANAGEMENT

STRATEGIC MANAGEMENT DECISION

PROJECT INCEPTION

DECISION TO BUILD
**DECISION ON THE STRATEGIC
DIRECTION OF PROJECT**

**POTENTIAL
TECHNIQUES**
- Strategic VM
- *Expert Choice*
- SNA (*Strategizer*)
- Brainstorming
- QFD
- SMART
- + OTHERS

Property
Solution

Organisational
Solution

Non-Building
Solution
- Outsourcing
- Subcontracting
- Downsizing

Decision to Build, Extend,
Lease, Renovate

**End of
Project
Inception**

Performance Brief

*Strategic Needs
Analysis* ends with
the production of the
performance brief

**Project Design
Stages**

Figure 6.3 Model of the project inception process: the decision to build.

stakeholders and other groups can understand and use as a base for the development of the project. The term *performance brief* is an emerging one where the focus of the document is on the *ends* rather than the *means*, with statements stakeholders can understand. Its emphasis is on the achievement of performance rather than the limited prescriptive approach that tends to inhibit innovation, effectiveness and efficiency. So, rather than state the area and types of spaces required, a performance brief identifies the output and level or standard required for various characteristics needed in the new project. In fact, some of these performance statements may eventually be satisfied in a form other than by a built solution or space. A different form of organisation (outsourcing or sub-contracting) or the introduction of new technology (greater automation or adoption of new forms of communications) may satisfy the performance requirement.

The performance approach has been promoted and encouraged by the International Council for Research and Innovation in Construction (CIB) through its Performance Based Building Thematic Network (*PeBBu*) in conferences and publications (CIB, 2003). The author has adopted the use of a performance approach at the strategic briefing stage and a number of other authors have espoused its use (Gray and Hughes, 2000; Jaggar *et al.*, 2002; Smith and Love, 2004; Smith, 2005a,b).

Table 6.2 Priority Levels and Performance Criteria.

Priority	Criteria
1. Key performance criteria	Service delivery
	Accessibility
	Security
	Council viability
	Effectiveness of individual service
2. Essential performance criteria	Interrelationship between Uses
	Flexibility
	Community ownership
	Sensitivity of urban design
3. Significant performance criteria	Diversity
	Sustainability
	Environmental efficiency
	Commercial viability
	Profile of building
	Extent of commercial uses

6.9   EXAMPLE PERFORMANCE BRIEF

In a workshop for a local government library project organised by Smith (2005a), the stakeholders developed and rated a range of 15 project concerns or requirements, prioritising them on a three-level scale of importance:

1. key performance criteria
2. essential performance criteria and
3. significant performance criteria.

The criteria are summarised in Table 6.2.

A sample of the first two key performance criteria, *Service delivery* and *Accessibility*, are shown in Appendix A where they are written as a performance brief. The performance brief was approved by the Glen Eira Council and it formed the basis for calling tenders for the design consultants. The project is now complete.

6.10   SUMMARY

The project inception stage has been the focus of intense research activity in design, project management and facilities management for a number of years. The need to establish the project parameters and performance requirements has been an imperative in many organisations, and facilities managers have been leaders in this process. Pre-design processes and activities are being instituted that work through client strategic and organisational issues, needs and requirements before the design team is involved (the business case). The participation of stakeholders in pre-design workshops is a common and welcome feature of many of these project inception approaches. These approaches prepare a clear and workable statement of the project requirements in performance terms that the client and user groups have committed themselves to. This strategic brief (or definition of the business case of the organisation) can then provide a sound basis for the documentation of the needs and provide a sound basis for the development of the design.

In the next chapter, Ornstein and Andrade describe a pre-design evaluation process that has the ability to receive, improve and execute the performance brief developed in the project inception stage, to provide a more effective and integrated facilities management solution to the organisation.

Finally, it is essential to guard against any project inception or pre-design process that merely creates the lowest common denominator result as a solution. These processes must follow a true problem-solving approach with no preconceptions of outcome, hidden agendas or cynical manipulation of stakeholders Most importantly, the process must aim to capture and encourage vision, leadership and creativity. All project inception approaches should aspire to this ideal. The processes must never become routine, formulaic and bureaucratic and a breeding ground for the average or the ordinary where innovation and creativity are stifled.

APPENDIX A: KEY PERFORMANCE CRITERION

Service Delivery (including effectiveness of individual service)

Goals

- To deliver Council services to the community in the most effective way in time and location.
- In the pursuit of joint use within the facility the quality and effectiveness of the individual Council services must not be reduced or compromised.

Key Objectives

Uses

1.1　The provision and transfer of existing uses in the proposed building are summarised in the table below.

EXISTING USES

- Library
- Youth services
- Playgroup
- Three-year old activity
- Neighbourhood program
- Senior citizens
- Open space
- Car parking

The design team should note that priority should be given to these existing uses in the new building.

1.2　The potential additional uses in the proposed building may include any of the uses summarised in the table below. In the potential uses, priority should be given where possible to those uses in the Council/Joint use category before moving into the commercial uses category.

Council/Joint

- Service centre
- Immunisation
- Meeting space
- Maternal and child health (Outreach)
- Education training rooms
- Information Glen Eira
- Library program and events
- Display area
- Toy library
- Library administration
- Glen Eira Adult Learning Centre
- Internet café
- Visiting services

Commercial

- Coffee shop
- Post office
- Business/commercial use
- Residential accommodation
- Business incubator

2.0   Hours of Operation

2.1   Flexible hours

Flexible hours of operation of the various services offered in the new facilities are a high priority. The use of the latest technology in achieving customer service at a time to suit demand is likely to play an important role in achieving this aim.

As an indication the existing uses are likely to have the following patterns of operation:

Service/use	Days	Times
Library	Monday to Friday	9.00 am to 7.00 pm
	Saturday	9.00 am to 5.00 pm
Youth Services	Monday to Friday	10.00 am to 8.00 pm
Playgroup	Monday to Friday	9.00 to 3.00 pm
Three-year-old activity	Monday to Friday	10.00 am to 2.00 pm
Neighbourhood program	Monday to Friday	3.00 to 8.00 pm

2.2   Hours of operation to suit users

The hours of operation must suit and be agreed with all user groups and designed to suit the customer demand in the catchment areas for each service. The potential for extending and changing these hours will depend on the opportunity for joint use with the additional uses identified in point 1.2 above.

3.0   Compatibility with other users

The type, extent and arrangement of uses and spaces must follow the principle of compatibility. All uses in the Centre must have a high level of compatibility with other existing and potential future uses.

Determining which uses shall be included will also depend on their complementarity with each other.

A final decision on the type and extent of the various uses will be made at the end of the Outline Proposals Stage when alternatives may be presented to the client group after a full investigation of the opportunities.

3.1 Timetabling

The joint use arrangements agreed between partners and service providers must be supported by effective timetabling arrangements through a formal management structure.

3.2 Compromise and conflict

Multi-purpose and joint use spaces must be designed and arranged to minimise conflict, reduce loss of effectiveness and promote higher levels of individual and joint service. All users must recognise that for joint use to succeed in multi-purpose areas will require service providers and to possibly make adjustments, compromise and concessions.

KEY PERFORMANCE CRITERION

Accessibility

Goal

• To provide easy and simple access to the Centre to all members of the community irrespective of their age, physical characteristics or means of transport.

Key Objectives

1. Prominence of Building

   The design of the building must emphasise the prominence of the building in the community with a strong image that encourages people to use its facilities and services.

2. Community Connections

   The Centre must be seen as a resource that is connected through its service provision to the community and externally to the wider neighbourhood of shops, services and to ratepayers.

   The Centre should improve accessibility through good transportation links with all forms of transport: cars, taxis, buses, trains, bicycles and pedestrians.

3. Acceptability

   It is essential that the new centre achieve a high acceptability and positive rating in the local community.

4. Easy Access

   The building must attain best practice standards for physical accessibility into, around and from the building for all groups: aged, disabled, children and all groups.

5. Quality of Facilities

   The quality and arrangement of facilities must aim to encourage a wider and more diverse range of use, and user, for those services.

6. Range of Services

   The range and mix of services in the new Centre, offered both immediately, and in the future, must be attractive to existing and potential users.

REFERENCES

Ackoff, R.L. (1978). *The Art of Problem Solving,* Wiley, New York.

Allinson, K. (1997). *Getting There By Design: An Architect's Guide to Design and Project Management.* Architectural Press, Butterworth-Heinemann, UK.

American Institute of Architects (AIA) (1987). *The Architect's Handbook of Professional Practice,* ed. Haviland, D.S., AIA, Washington, USA.

Ashworth, A. (2008). *Pre-Contract Studies: Development Economics, Tendering and Estimating,* 3rd edition, Wiley-Blackwell, Oxford, UK.

Atkin, B. and Flanagan, R. (1995). *Improving Value for Money in Construction: Guidance for Chartered Surveyors and their Clients,* Royal Institution of Chartered Surveyors, London.

Australian Institute of Project Management/Construction Industry Development Agency (1995). *Construction Industry Project Management Guide for Project Management Principals (Sponsors/Clients/Owners), Project Managers, Designers and Constructors,* Australian Institute of Project Management, Canberra.

Best, R. and de Valence, G. (Eds.) (1999). *Building in Value: Pre-design Issues,* Arnold, London.

Bilello, J. (1993). *Deciding to Build: University Organisation and the Design of Academic Buildings,* Unpublished PhD Dissertation, Faculty of the Graduate School, The University of Maryland, USA.

British Property Federation (BPF) (1983). *Manual of the BPF System,* British Property Federation, London.

Bromiley, P. (1982). *A Behavioural Investigation of Corporate Capital Investment,* Unpublished PhD dissertation, School of Urban and Public Affairs, Carnegie-Mellon University, USA.

Chiu, M.L. (1991). *Office Investment Decision-Making and Building Performance,* Unpublished PhD Dissertation, Department of Architecture, Carnegie-Mellon University, USA.

CIB (2003). *Performance Based Building: First International State-of-the-Art Report,* Prepared by Lee, A. and Barrett, P. for the CIB Development Foundation: PeBBu Thematic Network, CIB, Rotterdam, Netherlands.

Construction Industry Development Agency (CIDA) (1993). *Construction Industry Project Initiation Guide for Project Sponsors, Clients and Owners,* Commonwealth of Australia, Canberra.

CSM (Chartered Surveyor Monthly) (1998a). 7(9), p. 5.

CSM (Chartered Surveyor Monthly) (1998b). 7(9), pp. 55–65.

Dalton, E.H. (1989). *Capital Budget Decision-Making: The Practice in Ecuadorian Municipalities and Models of Explanation,* Unpublished PhD thesis, Syracuse University, USA.

Department of Industry, Science and Resources (DIST) (1999). *Building for Growth: An Analysis of the Australian Building and Construction Industries (Competitive Australia),* Commonwealth of Australia, Canberra.

Department of Health and Social Services (DHSS) (1986). *Capricode: Health Building Procedures,* HMSO, London.

Drage, J. (1970). Project coordination catches on. *Municipal Journal,* 78, 305–308.

Egan, J. (1998). *Rethinking Construction,* Construction Task Force Report, Department of the Environment, Transport and Regions, HMSO, London.

Gray, C. and Hughes, W.P. (2000). *Building Design Management.* Butterworth-Heinemann, London.

Hughes, W.P. (1989). *Organisational Analysis of Building Projects,* PhD Thesis, Liverpool Polytechnic (now Liverpool John Moores University), Liverpool, UK.

Hughes, W.P. and Walker, A. (1988). The organisation of public sector building projects. *Building Technology and Management,* August/September, 29–30.

Jaggar, D., Ross, A., Smith, J. and Love, P.E.D. (2002). *Building Design Cost Management,* Blackwell Science Publications, Oxford, UK.

Karma J.M. and Anumba C.J. (2001). A critical appraisal of the briefing process in construction. *Journal of Construction Research,* 2(1), 13–24.

Keel, D. and Douglas, I. (1994). *Client's Value Systems: A Scoping Study,* The Royal Institution of Chartered Surveyors, London.

Kirkham, R. (2007). *Ferry and Brandon's Cost Planning,* 8th edition, Wiley-Blackwell, Oxford, UK.

Latham, M. (1994). *Constructing the Team; Joint Review of Procurement and Contractual Arrangements in the United Kingdom Construction Industry,* Final Report, HMSO, UK.

Lawson, B. (1995). *How Designers Think,* The Architectural Press, London.

Ministry of Public Buildings and Works, Research and Development (1970) *The Building Process: A Case Study from Marks and Spencer Limited,* R&D Bulletin, HMSO, UK.

Mohsini, R.A. (1996). Strategic design: Front end incubation of buildings, in *North Meets South: Developing Ideas*, ed. Taylor, R.G.,Proceedings of CIB W92 – Procurement Systems, Department of Property Development and Construction Economics, University of Natal, Durban, South Africa, pp. 382–396.

Morton, R. and Jaggar, D. (1995). *Design and the Economics of Building*, Spon, London.

National Public Works Conference/National Building Construction Council Joint Working Party (NPWC/ NBC) (1990). *No Dispute*, Canberra, Australia.

Palmer, M.A. (1981). *The Architect's Guide to Facility Programming*, The American Institute of Architects, Washington DC, and Architectural Record Books, New York.

The Property Services Agency (1981), *Cost Planning and Computers*, Department of the Environment, London.

Rickards, T. (1990). *Creativity and Problem Solving at Work*, Gower, Aldershot, UK.

Royal Australian Institute of Architects (1993). *Client and Architects Agreement 1993*, Red Hill, Canberra.

Royal Institution of British Architects (1998). *Architect's Handbook of Practice Management*, 6th edn, RIBA Publications, London.

Royal Institution of British Architects (2000). *Architect's Plan of Work*, RIBA Publications, London.

Seeley, I.H. (1996). *Building Economics*, Macmillan, Basingstoke, UK.

Smith, J. and Jaggar, D. (2007). *Building Design Cost Planning for the Design Team*, 2nd edn, Elsevier, Oxford, UK.

Smith, J. and Love, P.E.D. (2004). Stakeholder management during project inception: Strategic needs analysis. *ASCE Journal of Architectural Engineering*, 10(1), 22–33.

Smith, J., Wyatt, R. and Jackson, N. (2003). A method for strategic client briefing. *Facilities*, 21(10), 203–211.

Smith, J. (2005a). Creating a user performance brief: An action research study, *11th Joint CIB International Symposium 2005, Combining Forces, Advancing Facilities Management and Construction through Innovation*, W55 Building Economics, W65 Organisation and Management of Construction, W60 Performance Concept in Building, W70 Facilities Management and Asset Maintenance, W82 Future Studies in Construction, Thematic Network PeBBu – Performance Based Building, 13–16 June 2005, Hall, Helsinki, Finland.

Smith, J. (2005b). An approach to developing a performance brief at the project inception stage. *Architectural Engineering and Design Management*, 1(1).

Walker, A. (2002). *Project Management in Construction*, 3rd edn, Blackwell Science, Oxford.

Walker, D.H.T. (1995). An investigation into construction time performance. *Construction Management and Economics*, 13, 263–274.

Weeks, C.W. (1980). *The Locus of a Select Non-Domestic Investment Decision: The Plant Location Real Estate Decision of US Based Non-Extractive Manufacturing Firms*, unpublished PhD Dissertation in Business Administration, Kent State University, Graduate School of Business Administration.

White, E.T. (1991a). *Facility Programming and the Corporate Architect*, Architectural Media, Tucson, Arizona, USA.

White, E.T. (1991b). *Project Programming: A Growing Architectural Service*, Architectural Media, Tucson, Arizona, USA.

Woodhead, R. (1999). *The Influence of Paradigms and Perspectives on the Decision to Build Undertaken by Large Experienced Clients of the UK Construction Industry*, unpublished PhD Thesis, School of Civil Engineering, University of Leeds, UK.

Woodhead, R.M. (2000). Investigation of the early stages of project formulation. *Facilities*, 18(13/14), 524–534.

Woodhead, R. and Smith, J. (2002). The decision to build and the organization. *Structural Survey*, 20(5), 189–198.

7   Pre-design Evaluation as a Strategic Tool for Facility Managers

Sheila Walbe Ornstein and Cláudia Andrade

CHAPTER OVERVIEW

This chapter sets out the principles of 'pre-design evaluation' (PDE) and illustrates how the methodological procedures in PDE can be used to inform the initial change management brief. Drawing on examples from the office environment, the chapter also shows the practical value of establishing the 'success' or 'failure' of existing facilities solutions before embarking on a change that results in a new facility.

Keywords: Pre-design evaluation; Office building; Workplace evaluation.

7.1   INTRODUCTION

Out of the three most important resources involving the management of a business — people, place and technology — there is no question that *place* is the most complex when it comes to investment decisions. Quick and effective decisions are required on all operating costs. These, in turn, arise from market dynamics — bringing about expansion and contraction. If a company occupying floors in a building needs to expand, it will have to rely on the availability of vacant spaces. Such space may not be available or may be located on floors remote from the occupied ones. This leads to an increase in travel distances and often a decrease in interaction. Either renting or building new spaces for the necessary expansion should be carefully considered, since it entails additional costs for an indefinite period of time with financial impacts on the organisation as a whole.

Therefore, the search for appropriate real estate and facilities solutions should consider space as an integral part of the business resource management process. The search will need to consider functional aspects such as comfort and security which can create value for the company and help to meet the overall corporate mission. Along these lines, McGregor and Then (1999) recognise that real estate and the supporting facility management function are critical strategic assets, building the interface between business units (satisfying competitive challenges, boosting productivity, gaining market share, minimising costs, maximising efficiency, amongst others) and the real estate/facilities function. This function involves monitoring the dynamics of the organisation to respond quickly to the business needs, as well as responding to unpredictable market conditions. This may include the installation of

Facilities Change Management. Edited by Edward Finch.
© 2012 Blackwell Publishing Ltd. Published 2012 by Blackwell Publishing Ltd.

new technologies and building systems to make the building more efficient and effective; increases in the quality of the work environment; reductions in operating costs and other innovations which will better support the financial, work environment and operational needs for the entire organisation.

In the Brazilian case, the pressure for cost reductions, rather than a focus on benefits, produces distortions that do not become evident due to the lack of integrated management among areas. The pressures facility managers are subjected to results in decisions based, in most cases, on the minimum value, without taking into account the analysis of how the decision will affect the performance of both the organisation and the people who work there.

7.2   THE PRE-DESIGN EVALUATION STAGE

The building characteristics are the result of decisions made at the time of the design process and should be originated at the pre-design (PD) stage which aims at accurately identifying the performance criteria that will support the building quality, optimising opportunities as well as correcting deficiencies that will better meet the expectations and needs of the users. Often in the beginning of this stage, the design of the intended building or refurbishment neither possess definite dimensions nor are their specific attributes well defined. Consequently, the number of possible alternatives are so diverse that it becomes essential to organise a preparatory strategic plan to define the direction to follow: whether to construct a new building or to select an existing one (new or otherwise); to buy or rent; or perhaps more simply to restructure/ reclassify/update the existing facility to better meet the needs of the organisation.

At this stage the focus is on the search for alternatives that may include:

1. the first ideas for a design
2. its general concept
3. the analyses of various scenarios and
4. the financial liabilities and associated deadlines.

During this search, some questions should be answered, from which the options and their impacts can be made clear so as to enable consistent decision making that will serve as a guide for the preparation of an effective brief or programming as it is named in the USA. Such a brief would vary according to the complexity of the design. However, its economic and financial viability will depend on aspects related to the surrounding location as well as those related to the building itself. The location will be influenced by the absence of risks such as crimes or floods for example, ease of access, the quality of the neighbourhood, existence of public transportation, commerce and services, among others. The quality of the building itself will determine comfort, internal security and functionality to better meet the needs of the organisation. A summary of the PDE objectives are expressed in Figure 7.1.

There is a broad range of situations in which PDE can be applied. This chapter intends to discuss both the application for existing buildings and/or new ones to be designed, as the methodological approach are very similar in both cases. The PDE itself should reliably convey the design intent as well as the size of investment required. Key questions are set out below.

• What is the current situation in the existing building?

The answer to such a question should address how the buildings and the surrounding environment is organised in relation to its occupancy (What benefit is being derived from its

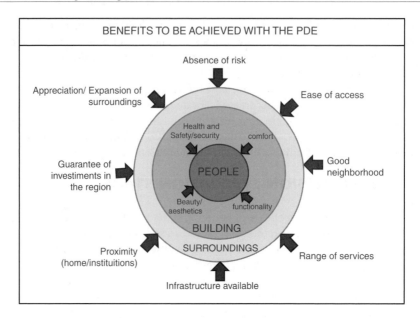

Figure 7.1 Benefits to be achieved through the pre-design evaluation.

occupancy?); functionality (How is the environment occupied?); occupant load (To what extent is the capacity of the building utilised?). Such a survey must also include an analysis of aspects related to the building's neighbourhood (location, risks, urban and service infrastructure; public transportation), as well as the qualities of the building's systems (air conditioning, lighting, elevators, fire prevention and fire fighting, information technology and electrical systems).

- How do people perceive the environments they currently use?

How environments meet people's needs for comfort, safety, health, functionality, flexibility, accessibility, sustainability, technology and how they influence their use, well-being and quality of life.

- What are the needs of users and other stakeholders (e.g. Heads of business units)?

The activities performed there; frequency of use; profile of occupants/users; legal requirements; and environmental quality.

- What are the required characteristics of the building/environment to be designed?

Independently of being a new building/environment design or a refurbishment of an existing one it is essential to take into consideration the following key aspects: the floor plan geometry, its dimensions and the façade characteristics in relation to the site and solar orientation, legal requirements, establishment of sustainable design requirements, flexibility to changes, internal and external aesthetic quality, symbolic function, technological support for development of the specific activities of the company as well as for the facility management of the work environment (Elmualim and Pelumi-Johnson, 2009) and its related costs. Lastly, PDE as well as post-occupancy evaluation (POE) (Preiser and Vischer, 2005; Ornstein and Ono, 2010) seek to evaluate the interdisciplinary approaches and the multidisciplinary teams for the development of the design, which constitutes the next step. Thus, defining what design team is the most appropriate, what profile is the most suitable for the design manager, and what disciplines/skills are required for its development are all key issues that will determine its quality and, consequently, the effectiveness of the whole process.

All these questions should be answered in the production stage of buildings/ environments prior to the development of the design and, in order to do so, procedures of environmental evaluation or PDE should be adopted. The greater the complexity of the building/environment that is to be changed/reclassified, the greater the need for a PDE, since it provides the briefing with consistent inputs. It thus impacts on the decisions made in the subsequent stages by designers, builders and facility managers. PDE provides the basis of a User Manual that outlines the activities and tasks of the facility manager, so it can have a significant impact on environmental performance during the use of the building. PDE can become a powerful tool for designers and other agents (stakeholders) that are active in the process of design, construction, use, operation, maintenance and management of built environments. This is because, as it analyses the design conditions beforehand, it provides clear guidelines on the blueprint to be developed for a productive work environment that properly meets the policies, organisational culture (van Meel, 2000) and human resources of the company. Fundamental to this is the choice between the individual cellular environment, controlled by individuals or by specific teams within the company (Appel-Meulenbroek, 2010) or, by contrast, by those elsewhere (Brunia and Hartjes-Gosselink, 2009). This has a significant impact on the future cost of occupancy (as well as effectiveness) in terms of workplace maintenance, whether the building is rented or owned. Therefore, it is crucial to address these alternatives and design guidelines (this being the essential role of a PDE), which must be clearly stated in the brief to be communicated not only to the contracting client, but also to future design teams. It can also be said that the costs of the implementation of a PDE compared with its benefits and the total costs of the undertaking are not very significant.

According to data from SINAENCO ((2008)) (Sindicato da Arquitetura e da Engenharia no Brasil; Brazil's Architecture and Engineering Syndicate), the design costs vary from 3.5% to 10% of the total cost of the construction of a building and the estimated overall cost of briefing represents up to 5% of the total project billing costs. Considering briefing as one of the pre-design activities, the total of a PDE will depend on the size and complexity of the design. However, it represents a low impact on the total value of the enterprise.

Internationally, there is for instance examples in countries such as the UK whose experience with public buildings shows that design costs (the full design process including the pre-design phase) are around 0.3–0.5% of the building whole-life budget and the construction costs are approximately 2–3% and that usually the costs of maintaining a public service is around 85% of the total and that design process quality could have a great impact on how well the building and its close neighbourhood perform during life time (NAO, CABE, OGC, Audit Commission, 2004).

Since the performance of the PDE stage is largely affected by the additional demands made in terms of programme time and additional personnel resources, some resistance may be encountered. However, in the absence of a clear course of action, such resistance may undermine decision making, causing rework throughout the process, construction delays and a higher final cost.

7.3   PRE-DESIGN EVALUATION: METHODS AND TECHNIQUES

Zeisel (2006) and Pinheiro and Günther (2008) systematised and made available with didactic examples, scientific research procedures that can be applied to both pre-design evaluations (PDEs) and post-occupancy evaluations (POEs). In line with this, it is noted that PDEs and POEs require very similar methodological procedures (Andrade, 2000; Federal Facilities Council, 2002; Andrade, 2005; Preiser and Vischer, 2005; Andrade, 2007; Ornstein and Ono, 2010), since a stage feeds the other in a continuous cycle.

Clements-Croome (2000) associates the results of pre-design performance evaluations in work environments in relation to productivity. Jong and Van der Voordt (2002) and Van der Voordt and Van Wegen (2005) point out that the PDE stage is an experiment or trial, which substantiates the suitability of employing the aforementioned 'fitness for purpose' test in this stage. Maarleveld *et al.* (2009) describe research on the application of the WODI (Work Environment Diagnostic Instrument) toolkit in order to measure user satisfaction in new offices. The authors discussed the employees' perception and expectation regarding issues such as the functionality of workplaces, location of worksites, privacy, storage facilities, facility management, perceived organisational productivity and so on.

Nelson (2006) more recently discusses PDE in the light of quality management and design standards in architecture and the implementation of ISO 9000 (2000). (International Organization for Standardization (ISO) is the norm related to systems of quality management.) According to this author it is under the design manager and the design team responsibilities to specify which standards should be considered as appropriate and how they could be applied. Also designers could bring to the 'standard process' the customer (user) standards. In other words, PDE probably is the most adequate phase to identify standards from both professional and client point of views.

PDE uses, in smaller or larger scales, methods, techniques and quantitative and qualitative tools that explore the emerging requirements of the occupants and the capacity of the building to meet them. To this end, a multi-method approach is critical to guarantee the consistency and credibility of all the data found (Preiser and Schramm, 2005). The methodological procedures to be adopted in a PDE depend on the size and complexity of the environments under analysis and also on the interventions or renovations to be carried out, the number of users (customers and end users) and activities (types and the way they are performed) that are to be housed in these environments. They also depend on the time available for the performance of the PDE stage, since it includes, in addition to surveys, interviews, application of questionnaires and *in situ* measurements, and the consolidation, processing and analyses of data for the making of an accurate and detailed diagnosis.

In general, since PDE has interdisciplinary characteristics and is part of the strategic planning that comprises the three vectors — people, processes and environment — of a company, it can be said that the more extensive and profound the changes in these vectors are, the greater the benefits of the investment in the application of a PDE.

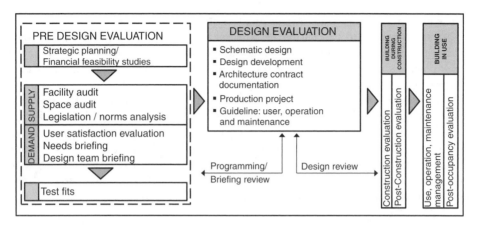

Figure 7.2 PDE flow chart.

The application of PDE methods and techniques, as listed in Figure 7.2, provides diagnoses and other inputs to answer these questions according to the following evaluation tools.

7.3.1 Facility audit (performance evaluation of the building infrastructure)

The facility audit aims at identifying the conservation status and efficiency of the current building systems and their infrastructure. This involves the assessment of needs for improvements and/or technology updates to better meet the requirements. These requirements include comfort, safety and security. At this stage it is also necessary to measure the investment required, time limits and the feasibility of this. This is a fundamental technique since '. . . financial constraints as well as time constraints are only issues that need to be clarified during strategic planning before facility programming can start.' (Schramm, 2005, p. 31).

For this purpose a team of consultants from different fields of expertise should be hired — air conditioning, electrical installations, security systems, automation, among others — to assess the existing situation through *in situ* surveys, technical analysis of the equipment and interviews with those responsible for the technical areas of the building (facility manager, maintenance manager, security, IT, among others).

7.3.2 Space audit (performance evaluation of the physical occupation)

A space audit involves evaluation of the level of efficiency and performance of the present building and its existing environments by means of quantitative analyses, inspections and measurements, assuming that the action of planning the construction and/or occupation of a building must be preceded by an understanding of the current situation. This method of space usage analysis is important because it enables decision makers to map problems and weaknesses, identify opportunities for improvement and the positive aspects to be kept in a new design. This is achieved through the understanding of *how* the physical space within the buildings is occupied, *how much* of it is occupied and *who* occupies it.

The procedures adopted involve an analysis of the amount of area occupied by each type of space usage and by each of the business units that make up the organisational structure that will occupy this space, previously identified and quantified by means of layout plans, in local surveys and meetings with head technicians and managers of each business unit.

In the case of office buildings, for example, how the organisational structure is distributed according to the organisational chart of the company and the distribution of the different types of space usage are evaluated. This is typically based on work environment design parameters such as the areas intended for workstations, meeting rooms, ancillary spaces, general storage and main circulation routes (to areas such as training centres, cafeterias, libraries and technical areas such as data centres and IT rooms, among others).

7.3.3 Survey and analysis of the legislative restrictions

This stage is crucial because it will determine the restrictions as well as the technical standards to be followed in the design at the time of making a decision. In preliminary stages, as in the strategic plan, this step can show the technical or financial infeasibility of an enterprise. That is why it is important to identify these requirements prior to the design. In addition to local legal aspects, it must be mentioned that many public/government institutions or private companies operating at a global level have their own standards and guidelines which must be included in this analysis. More and more large enterprises (road infrastructure, hospitals, schools, etc.) have also developed their own rules as is the case with Public-Private Partnership projects.

7.3.4 Financial feasibility studies

This procedure involves an assessment of the financial aspects involved in a design. Financial viability is paramount. The strategic definition involves consideration of construction or renovating activities and move activities. As such, these activities should be preceded by a careful analysis of all the costs involved: rent, condo fees, operating costs of the building, financial costs resulting from contract terminations, relocation and surrendering costs, costs related to the payment of design fees, architectural interventions, infrastructure adaptations, purchase of furniture and equipment. The costs associated with the disruption to productivity arising from a change of location should also be considered. For more complex processes, the hiring of and/or the involvement of a specialised professional is recommended so that the cost factor is weighed against the benefits to be achieved, thus resulting in greater certainty at the time of decision making.

7.3.5 User satisfaction evaluation

This set of methods concerns the evaluation of how users perceive the environment they use based on questionnaire surveys, identifying the strengths and weaknesses related to the environment in general, and offers ideas for further improvement.

Questionnaires, after being pre-tested and adjusted if necessary, can be applied to a statistically representative sample of the population that occupies the building (most questions are associated with ranges of values). Questionnaires can be applied face to face or even via a corporate intranet which can save time through automatic data capture and storage. They should be reader friendly, easy to be filled out and the respondent should be able to complete it quickly in order to avoid a long absence from their work activities.

As to office buildings, due to numerous variables related to comfort, health and functionality, it is also recommended that questionnaires be proportionally distributed to the different floors and/or workplace. This stratified sample also need to be based on solar orientation, workplace position in relation to the floor layout (near to or surrounding core areas, windows, etc.), gender and age. This procedure will enable the division of the sample into groups and the performance of spot measurements exactly where the problems have been detected if, for example, those related to environmental comfort.

In some situations it is possible to opt for the implementation of focus groups composed of various profiles of users/occupants. Such a sample should not exceed 15 participants in each group, according to what has been determined by Zeisel (2006), so that the interviewer/moderator has control over all participants and these can contribute in a more effective manner.

In cases where there is a lot of data to be collected, and to enable subsequent comparison among groups, it is recommended that the opportunity is given for people to reflect on their environments and engage in the articulation of individual satisfaction levels. In this way there is an opportunity for discussion and affords the possibility of delving into specific points (Zeisel, 2006).

7.3.6 Data gathering for the project briefing

This stage is predominantly a qualitative evaluation involving numerous sources. It involves the collection of information through: (1) interviews, (2) focus groups, (3) walkthroughs with checklists, (4) observations of activities and behaviours, (5) inspections and measurements of people's needs in relation to their physical space, as well as through conversations with those who will manage, operate, maintain and conserve the building during its use.

Such methodologies apply techniques which are consistent with the goals of the study.

1. Interviews: these provide a qualitative tool for data collection, including the viewpoints of key people, namely the company's chief executive, director, general managers, facility managers and others. As a rule, semi-structured interviews drawing on a previously defined template script that includes some topics or questions are adopted which allow for requisite coverage of issues. They can be recorded (if the interviewee agrees) or written down by the interviewer for a subsequent analysis. It is suggested to have them done such that the interviews should be scheduled to last less than an hour. Interviews provide the starting point for the preparation of the questionnaire to be distributed to users.

2. Focus groups: these provide a qualitative tool for eliciting viewpoints of stakeholders, including directors, managers or technicians. As with interviews, it also relies on a script of topics and questions that serve as a guide for the moderator who asks questions, elicits comments and also redirects conversations if they stray from the object of the focus group. The moderator should preferably be supported by an assistant who takes notes on the impressions expressed by participants during the activity, controls the recording device and takes pictures (if participants agree to either of these forms of recording). The number of participants should be between five and seven, but in some cases, as pointed out before, it can be up to 15, in addition to the

moderator and his or her assistant. Each activity usually takes between 30 and 60 minutes. Focus groups can help in understanding answers written on questionnaires and, in instances where there is no time for the use of a questionnaires, focus groups provide a flexible and more timely alternative.

3. Walkthroughs with checklists: this tool involves a planned tour of the environment under study, assessed by the hired evaluator, where the object is the currently occupied environment. The walkthrough provides an understanding of the conditions of the space and the existing infrastructure as well as the corresponding conservation status. This allows the analysis of options regarding either reuse or relocation. The evaluator may carry out this activity with key people (the facility manager, the technician responsible for air conditioning, the human resources director, the professional in charge of information technology) who will also be encouraged to point out specific issues. To complete this task, the evaluator should possess the as-built plan and layout for notes and as a checklist of the physical aspects (comprising, for example, building system items, furniture, environmental conditions, among others). He or she must also bring a tape measure and a camera. All the information should be written down on a tailored proforma.

4. Observation of activities: it is still possible to map individual and group activities (including those at workstations and during circulation and interaction). This mapping can be supplemented by measurements of flows in circulation areas and entrance and exit areas. Mapping can also be used to understand the dynamics of space usage with the intention of optimising and use of space sharing. The mapping should sample a typical week. The weekend should be included if there are activities during this period. For the application of this tool, it is necessary to have the as-built plan and layout, know the locations where people work and the activities carried out.

5. Inspections and measurements: these tools can be applied in environments where interventions are likely to occur (allowing a 'before and after' assessment). They include the assessment of fire safety conditions, use, property safety, accessibility, ergonomic and functional aspects of furniture (workstations, meeting rooms, recreational areas and others). Measurement of environmental comfort conditions may be carried out by using portable devices that evaluate thermal comfort, humidity, ventilation, lighting comfort (natural and artificial lighting), acoustic comfort and other related factors. Ergonomic aspects related to work are also evaluated through observations and measurements of items such as chairs and tables. In the evaluation of 'fitness for purpose', a test fit of the space plan can also be used to assess existing or intended replacement facilities as explained below. (A test fit is a quick rough preliminary study just to show that the space is sufficient to attend the design brief (Rayfield, 1994)).

7.3.7 Design team briefing (focus on the definition of the design team)

This stage draws on the analysis and consolidation of all the data. A consistent diagnosis is made to serve as the basis for the specifics of the design brief. This will contain all the information needed to hire the team of designers and consultants, including the identification of the designers' team coordinator profile, for the development of each design, to have it approved at all levels, and for its subsequent implementation, according to deadlines, costs and quality requirements set in advance.

7.3.8 PDE final report

According to Moreira and Kowaltowski (2009), the requirements schedule should gather information on the characteristics and requirements of the client, end users and the context (indices, standards, legislation and benchmarks), taking into account documents from various sources and of different types. Ideally these should be expressed diagrammatically or as charts, demonstrating clearly the objectives of the design of the building and/or its design interiors. Thus the requirements schedule includes results from all tools applied during PDE, from the facility/space audit to the design team briefing. It defines priorities, weights and attributes in relation to specific issues and guidelines. This in turn will assist with defining the functional requirements of the future environment. Cherry (2001) also stresses the importance of the needs or architectural programme as a procedure to promptly recognise who the client(s) is/are, the needs that they have and the likelihood of meeting them.

Thus PDE analyses what is right in a new design and should be kept, what is wrong and needs to be corrected, the opportunities to optimise the environment and the needs to be addressed in a new building or environment. This enables the creation of comprehensive guidelines to be followed by the design team and monitored by the management team.

From these guidelines, it is possible, in addition to the development of a design, to develop a strategic occupation plan for the building or environment to be constructed, establishing the parameters, the concept of occupation and the distribution of groups in the building (even at the level of individual floors) which will facilitate the subsequent process of developing interior designs and layouts.

Lately, 'fitness for purpose' simulations (3D simulations, workplace prototypes and so on) have become common when dealing with large processes involving changes of location or more complex restructuring. The client can thus see beforehand that the proposed environment will actually abide by the defined brief and will also provide a preview of what the environment to be designed may look like. As such, it serves as a framework for choosing the typology of the building or the building to be leased/purchased.

Thus this chapter shows how the use of methods and techniques of PDE can aid clients who need to make decisions about moving to a new office building or making adjustments to existing work environments. It enables them to identify the best options through low-cost processes that bring significant benefits, taking into account the total financial impacts of these initiatives on the budget of the organisation.

The international literature shows that real estate enterprises such as office buildings or work environments incur many risks. These arise long before the start of the conventional design activities (preliminary study, schematic design, basic design, executive design and, more recently, design for production) (Atkins and Simpson, 2008). The management and reduction of these risks result in the fulfilment of expectations of both the client and end users as well as the financial constraints. The designers and the potential design coordinator should take part in the PDE stage. The architect's controlling force can often be compromised by real estate market agents, particularly those involved in marketing, development and brokerage, taking over the tasks of formulating the briefing and the programming. The PDE provides a mechanism for maintaining the visibility of the architect in this respect.

The American Institute of Architects (2002) makes evident the importance of this stage in which the architect and team assist the client in clarifying the goals to be achieved and the guidelines of the design and its limitations. In this stage, data on people, legal aspects, equipment and other issues are collected to be analysed and presented in diagnostics. During the PDE stage, the analysis of land options or options for existing environments is also accomplished by means of checklists.

Preiser (1993) stated in a pioneering way how relevant the activities of pre-design are to both the teaching of architecture and the professional practice, particularly facilities management. Investigations undertaken by means of proper methods, techniques and tools can determine the space planning, building infrastructure, furniture, facilities and services requirements. It can also address specific aspects such as the image of the building and the company, layout and budget constraints to meet the client's needs. Hershberger (1999) highlights in detail the importance of the briefing (programming) stage and the architects role in this briefing. They are not simply concerned with checklists and room dimensions. They also have a key role to play in specifying the relationships among environments and defining layouts, facilities and infrastructures according to the business profile of the organisation.

7.4   CASE STUDY EXAMPLE

This case study concerns a multinational company of European origin in the industrial food sector with approximately 1187 employees that occupies a building with 18 floors, about 1100 square meters each, built in the late 1980s and located in the South Eastern part of the City of São Paulo, Brazil.

This PDE was undertaken in two stages: the first consisted of a plan whose objective was to define whether the company should continue in the same building or move into another. The capacity of the building to meet the needs of the company was compromised, requiring a number of interventions for its updating with regard to both its infrastructure (replacement of air conditioning equipment, improved lighting systems, expansion of the supply of energy, replacement and expansion of the electricity grid, of voice communication and data, among others); and the capacity of the building to meet the needs of its users (complete renovation of the workplace, including the implementation of a new workplace concept, more in line with the dynamics of the organisation). To add an extra element to the issue, the value of rent was high compared to current real estate market numbers for buildings of the same category at that time.

The first step was to undertake an accurate diagnosis of the present situation, made by a team of specialists, coordinated by the facility manager of the company. This followed the implementation of local surveys and measurements. At the same time, the professionals in charge of technical and safety issues were interviewed. Planning questionnaires were sent to the managers of every business unit so that they could report possible increases in headcount. Specific needs needed to be identified when evaluating the size of the building that should be sought in the real estate market.

This diagnosis enabled the quantification of the floor area needed, bearing in mind that some of the existing spaces in the present building should not necessarily exist in a new one, such as the restaurant (too large and replaceable if meal tickets are made available). This resulted in the preparation of a briefing with all basic requirements divided into four key points, and used by the real estate consultancy service hired by the company as a parameter for the search and selection of buildings.

1. Location in the city and characteristics of the surroundings (accessibility by public transportation, good infrastructure of services such as bank branches, restaurants, shopping centres, etc); absence of property hazards (floods, for instance) and low crime rate; analysis of technical issues.
2. Characteristics of the building, including the number of parking spaces and its infrastructure.
3. Efficiency of the standard floor type (size, characteristics, number of restrooms, kitchens, ceiling height, etc.).
4. Last, but not least, the financial indices related to leases and operating costs.

The four buildings presented were surveyed by every participant in the process and a checklist, containing all the information about the four key points mentioned above, was filled out. At the same time, the efficiency of the standard floor type was evaluated, generating a spreadsheet that provided the data used for the disqualification of two buildings: one for failing to meet the basic technical requirements related to typology, size and floor efficiency and the other for being in an inconvenient location (see Figures 7.3 and 7.4).

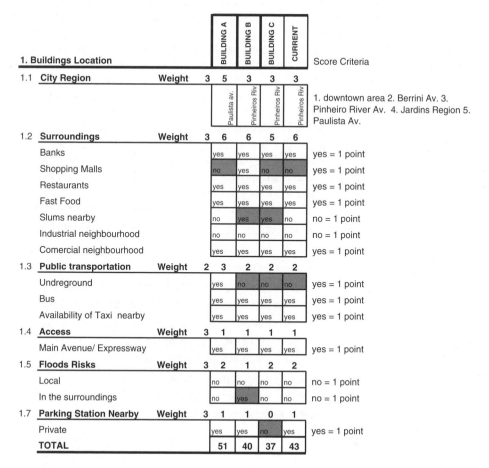

1. Buildings Location		BUILDING A	BUILDING B	BUILDING C	CURRENT	Score Criteria
1.1 **City Region**	Weight 3	5	3	3	3	1. downtown area 2. Berrini Av. 3. Pinheiro River Av. 4. Jardins Region 5. Paulista Av.
		Paulista av.	Pinheiros Riv	Pinheiros Riv	Pinheiros Riv	
1.2 **Surroundings**	Weight 3	6	6	5	6	
Banks		yes	yes	yes	yes	yes = 1 point
Shopping Malls		no	yes	no	no	yes = 1 point
Restaurants		yes	yes	yes	yes	yes = 1 point
Fast Food		yes	yes	yes	yes	yes = 1 point
Slums nearby		no	yes	yes	no	no = 1 point
Industrial neighbourhood		no	no	no	no	no = 1 point
Comercial neighbourhood		yes	yes	yes	yes	yes = 1 point
1.3 **Public transportation**	Weight 2	3	2	2	2	
Undreground		yes	no	no	no	yes = 1 point
Bus		yes	yes	yes	yes	yes = 1 point
Availability of Taxi nearby		yes	yes	yes	yes	yes = 1 point
1.4 **Access**	Weight 3	1	1	1	1	
Main Avenue/ Expressway		yes	yes	yes	yes	yes = 1 point
1.5 **Floods Risks**	Weight 3	2	1	2	2	
Local		no	no	no	no	no = 1 point
In the surroundings		no	yes	no	no	no = 1 point
1.7 **Parking Station Nearby**	Weight 3	1	1	0	1	
Private		yes	yes	no	yes	yes = 1 point
TOTAL		51	40	37	43	

Figure 7.3 Partially filled checklist with the results of walkthrough. Compiled from Andrade Azevedo Arquitetura Corporativa databank.

TOTAL SCORE	BUILDING A	BUILDING B	BUILDING C	CURRENT
Building Location and Surrounding	51	40	37	43
Techinical Issues	74	92	84	72
Financial Issues	57	69	60	77
TOTAL	**182**	**201**	**180**	**192**

Figure 7.4 Total score obtained for the four buildings. Compiled from Andrade Azevedo Arquitetura Corporativa databank.

Next, the requirements needed for the move were studied, the impacts assessed and the size of the investment determined. Consideration was given to the need to adapt each of the buildings, including the currently leased one. The newly determined needs of the company were measured in financial terms. The resulting data was used as the basis for the negotiation between the hired real estate consultancy firm and the landlord of the present building. Eventually this produced very positive results for the tenants, who then decided to remain in the current building and make the necessary adjustments to fulfil new needs. From there the second step of the work was initiated.

This second step dealt with the determination of the requirements for developing interior and technical designs; the specification of the needs programme and the preparation of a new concept of occupation that could better cope with the dynamics of the work in an effective manner in terms of safety, comfort and functionality and that, at the same time, would be stimulating for the staff. To this end, the following methods and evaluation techniques were used: space audit for the analysis of the efficiency of the current workplace layouts, application of questionnaires survey to users to find out about their perception level in relation to their work environment (see Figure 7.5), interviews with the managers of each area for the specification of the needs programme, surveys, local observations and ergonomic-functional analyses of the existing furniture.

The analysis of all this information resulted in a more precise diagnosis of the occupation of the building, the problems identified and, above all, of their impacts on the satisfaction

CASE STUDY

coffee area	39%
storage	45%
accoustic levels	48%
workstation	53%
toilets	58%
temperature	62%
meeting rooms	71%
privacy	78%

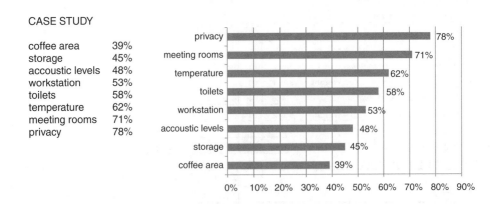

Figure 7.5 Results of the users survey: items to be improved. Compiled from Andrade Azevedo Arquitetura Corporativa databank.

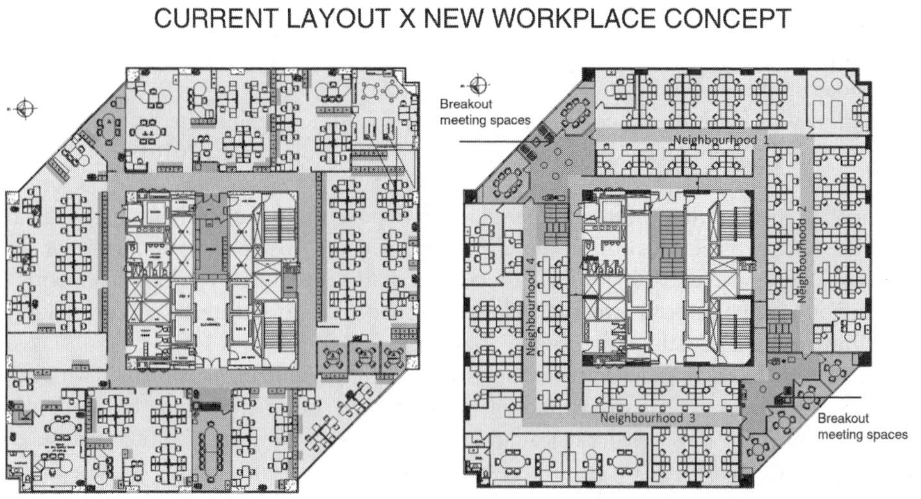

CURRENT LAYOUT X NEW WORKPLACE CONCEPT

Figure 7.6 Example of the current workplace layout floor plan and the new workplace concept floor plan (test fit). Compiled from Andrade Azevedo Arquitetura Corporativa databank.

and performance of the workers. This was in addition to the fine tuning of the finance and the preparation of a detailed briefing that guided all the designs.

It should also be added that all this documentation was presented to the board of the organisation and that its immediate approval took place. This can be explained by the payback analysis that was undertaken that showed in a clear, concise and consistent manner tangible benefits such as the elimination of moving costs, the absorption of the necessary additional staff on account of a more optimised concept and the improvement of the physical conditions of the building. In addition, intangible benefits were identified such as the increased productivity due to the improvement of the workplace conditions by a new space plan concept that better support the user's needs (Figure 7.6).

7.5   CONCLUSIONS

The most efficient way to start a design process is to have undertaken a pre-design evaluation (PDE). This is particularly so in the case of large and/or very complex work environments. This PDE should clearly present the results of the research carried out, the regulatory and legal constraints that might exist in the design guidelines and should also try to set benchmarks.

Concise, reader friendly and easy-to-understand schemes, charts and tables, in addition to the definition of benchmarks against which analyses can be made, are welcome at this stage. They provide support to decision makers, not only in relation to the design to be developed, but also in relation to the design team to be hired.

As we have seen, PDE can help the entrepreneur and designers in steering the development of the design. The PDE steps provide the basis for more consistent and grounded proposals in terms of:

1. the profile of the companies
2. the location/geography

3. climate
4. culture
5. socioeconomic conditions
6. relationships between the environment and behaviour set against the financial performance of the company (Appel-Mewlenbroek, 2010; Brown *et al.*, 2010).

The PDE methodology bears some similarities to post occupancy evaluations, which are concerned with the evaluation of building performance post occupation (Preiser and Vischer, 2005), particularly if the objective is to reclassify or renovate work environments in use; and one evaluation can give feedback to the other in case of future designs or buildings with similar characteristics (see Figure 7.2 above).

PDE should be encouraged in enterprises in developing countries like Brazil, in which the strategic planning and its corresponding tools such as the briefing and the needs programme, if properly developed, can lead to a reduction in costs during all subsequent stages, with an increase in the well-being of end users, better environmental quality (sustainability) and a longer lifecycle, and with lower costs of corrective maintenance, also meaning an optimised cost—benefit relationship in the case of other parties such as developers, builders and facility managers.

Although the technique of PDE is carried out in developed countries and is also taught in undergraduate courses in architecture, urbanism and engineering, PDE in developing countries is rarely explored and presented to the future professionals as a set of systematic procedures that can assist in professional practice. Not even in countries such as Brazil is there a representative group of clients (entrepreneurs or developers) that recognise the stage prior to the development of a design as a justifiable activity. The technique can bring great benefits in the early stages of design, since it is then that all the positive and negative aspects of a layout or renovation of a work environment are identified. The design and construction phases continue to attract the greatest attention because of the tangible nature of the final 'product'. Some of the difficulties faced by facility managers in the daily operation and maintenance of work environments may derive from the absence or deficiency of a PDE. We hope this chapter can raise awareness of the importance of PDE in the design process and in the performance of buildings in use.

REFERENCES

American Institute of Architects (2002). *The Architect's Handbook of Professional Practice* (Joseph A. Demkin, executive editor) John Wiley & Sons [Student edition], New York.

Andrade, C.M. de (2000). *Avaliação da Ocupação Física em Edifícios de escritórios utilizando métodos quali-quantitativos. O caso da Editora Abril em São Paulo.* Faculdade de Arquitetura e Urbanismo da Universidade de São Paulo (unpublished Masters thesis), São Paulo.

Andrade, C. M. de (2005). *Avaliação de Desempenho em Edifícios de Escritórios: o ambiente de trabalho como meio para o bem estar produtivo.* Faculdade de Arquitetura e Urbanismo da Universidade de São Paulo (unpublished Doctorate dissertation), São Paulo.

Andrade, C.M. de (2007). Edifícios de escritório: avaliação de desempenho do ambiente produtivo. *Tempo, Cidade e Arquitetura (Arquiteses 1).* Annablume, Faculdade de Arquitetura e Urbanismo da Universidade de São Paulo, Fundação para a Pesquisa Ambiental, pp. 341—358.

Appel-Meulenbroek, R. (2010). Knowledge sharing through co-presence: added value of facilities. *Facilities. Environmental Behaviour in Facilities Management*, 6(3;4), 189—205.

Atkins, J.B. and Simpson, G.A. (2008). *Managing Project Risk. Best Practices for Architects and Related Professionals.* John Wiley & Sons, Inc., New Jersey, USA.

Bogers, T., Van Mell, J. and Van der Voordt, T. (2008). Architects about brief. *Facilities*, 26(3/4), 109–116.

Brown, Z., Cole, R.J., Robinson, J. and Dowlatabadi, H. (2010). Evaluating user experience in green buildings in relation to workplace culture and context. *Facilities. Environment Behaviour in Facilities Management*, 28(3/4), 225–238.

Brunia, S. and Hartjes-Gosselink, A. (2009). Personalization in non-territorial offices: A study of a human need. *Journal of Corporate Real Estate*, 11(3), 169–182.

Cherry, E. (2001). Orchestrating client involvement, in *Architectural Design Portable Handbook* (Andy Pressman). McGraw Hill, New York, pp. 90–98.

Clements-Croome, D. (ed.) (2000). *Creating the Productive Workplace*. Spon, UK.

Elmualim, A. and Pelumi-Johnson, A. (2009). Application of computer-aided facilities management (CAFM) for intelligent buildings operation. *Facilities. Intelligent Buildings and Their Operation Within the FM Context: Concepts, Opportunities and Beneficiaries*. 27(11/12), 421–428.

Federal Facilities Council (2002). *Learning from Our Buildings. A State-of-the-Practice Summary of Post-Occupancy Evaluation*. National Academy Press, Washington DC, USA.

Hershberger, R.G. (1999). *Architectural Programming and Predesign Manager*. McGraw Hill, New York.

ISO – International Organization for, Standardization. (2000). *ISO 15686-1. Building and Constructed Assets – Service Life Planning – Part 1: General Principles*. London.

de Jong, T.M. and Van der Voordt, T.J.M. (2002). *Ways to Study and Research Urban, Architectural and Technical Design*. Delft University Press, Delft, Netherlands.

Maarleveld, M., Volker, L. and Van der Voordt, T.J.M. (2009). Measuring employee satisfaction in new offices – the WODI toolkit. *Journal of Facility Management*, 7(3), 181–197.

McGregor, W. and Then, D.S. (1999). *Facilities Management and the Business of Space*. London, UK: Arnold Publishers.

van Meel, J. (2000). *The European Office. Office Design and Nation Context*. 010 Publishers, Rotterdam, Holland.

Moreira, D. de C. and Kowaltowski, D.C.C.K. (2009). Discussão sobre a importância do programa de necessidades no processo de projeto em arquitetura. *Ambiente Construído.9(2)*. Porto Alegre: Associação Nacional de Tecnologia do Ambiente Construído, 31–45.

National Audit Office (NAO), Commission for Architecture and the Built Environment (CABE), Office of Government Commerce (OGC), Audit Commission (2004). *Getting Value for Money from Construction Projects through Design. How Auditors Can Help*. CABE, UK. [available at www.cabe.org.uk, accessed July 2010].

Pinheiro, J. de Q. and Günther, H (2008). *Métodos de Pesquisa nos Estudos Pessoa – Ambiente*. Casa do Psicólogo, São Paulo.

Nelson, C. (2006). *Managing Quality in Architecture. A Handbook for Creators of the Built Environment*. Elsevier, UK.

Ornstein, S.W. and Ono, R. (2010). Post-occupancy evaluation and design quality in Brazil: Concepts, approaches and an example of application. *Architectural Engineering and Design Management*, 6, 48–67.

Preiser, W.F.E. (1993) *Professional Practice in Facility Programming*. Van Nostrand Reinhold, New York.

Preiser, W.F.E. and Vischer, J.C. (eds.) (2005). *Assessing Building Perfomance*. Elsevier, UK.

Preiser, W.F.E. and Schramm, U. (2005). A conceptual framework for building performance evaluation. In Preiser, W.F.E. and Vischer, J.C. (eds.), *Assessing Building Performance*. Elsevier, UK, pp. 15–26.

Rayfield, J.K. (1994). *The Office Interior Design Guide*. John Wiley & Sons, New York.

Schramm, U. (2005). Phase 1: Strategic planning – effectiveness review. In Preiser, W.F.E. and Vischer, J.C. (eds.), *Assessing Building Performance*. Elsevier, UK, pp. 29–38.

SINAENCO – Sindicato da Arquitetura e da Engenharia (2008). *Roteiro de Preços: Orientação para Composição de Preços de Estudos e Projetos de Arquitetura e Engenharia. Anexo 4*. www.sinaenco.com.br [accessed in October 2010].

Van der Voordt, T.J.M. and Van Wegen, H.B.R. (2005). *Architecture In Use. An Introduction to the Programming, Design and Evaluation of Buildings*. Elsevier, Oxford, UK.

Zeisel, J. (2006). *Inquiry by Design. Environment/Behavior/Neuroscience in Architecture, Interiors, Landscape, and Planning*. W.W. Norton & Company, New York.

8   Implementing Change

Melanie Bull and Tim Brown

CHAPTER OVERVIEW

Facilities management is very much about people, and this can sometimes be forgotten. Facilities management is the enabler for any organisation, and there is a need to engage with end users to ensure the service we are offering allows them to carry out their day to day business. When we start to disrupt this service, even if it is actually for the better, it can still be very emotive.

This chapter addresses the issues that surround the implementation stage of change projects from an FM perspective, considering the communications issues of move management, the practical issues involved and methods to minimise the disruption associated with a move. It draws on a practical case study example to illustrate the facets of communication in move management. The issues arising from changes to working environments can be hard for facilities management staff to engage with. There is very often an approach that 'it just has to be done' and this results in a lack of engagement and communication from the facilities management staff to the end users (Donald, 1994; La Framboise *et al.*, 2003; Price and Fortune, 2008). There is existing literature in relation to the importance of change communication and how the lack of a communication strategy can impact on the satisfaction and engagement of staff in the long term, but little is written about this from a facilities management perspective. Included within the chapter is a case study based on research in a blue chip organisation that focused on staff satisfaction following a change to working practice and also on the communications method used. Alongside the issue of communication, the chapter also considers the politics of move management, with recommendations on how to engage staff so they are fully participatory in the move, building the right project team and how to evaluate the communication strategy used and the overall satisfaction of the staff after the move.

Keywords: Communication; Change; Moves; Satisfaction; Projects.

8.1   PARTICIPATION IN THE MOVE

Campbell and Finch's (2004) paper on customer satisfaction and organisational justice explored the concept of justice in relation to facilities management and their customers. The key themes emerging from this paper, within procedural justice in service provision

Facilities Change Management. Edited by Edward Finch.
© 2012 Blackwell Publishing Ltd. Published 2012 by Blackwell Publishing Ltd.

was the question of whether facilities management needed to engage customers in decision making, to allow them to feel they have had a voice and also to be able to understand their organisational needs with the aim of improving relationships and buy in to changes. At a time of change, there can be a tendency to feel that the facilities management team is imposing change on staff as opposed to it being an organisational need. The concept that underpins organisational justice is an idea of fairness; how do we engage staff in open and honest two-way communication in relation to a facilities management change?

Distributive justice (Homans, 1961) refers to perceived fairness of outcomes, whereas procedural justice could be considered to focus more on the mechanisms and processes used in order to achieve the outcomes (Folger and Cropanzano, 1998). For facilities management, it is usually the procedures that fall short in relation to gaining buy-in from staff and the case study included within this chapter highlights this point. So how can we ensure our procedures are fair to the end users? Is the idea of fairness a subjective point? To some extent there is a level of subjectivity with anything that relates to customer satisfaction: one person's excellent is another's mediocre! Although in the case of procedural justice perhaps it is being able to understand the customer that we are communicating with and trying to build relationships and trust to reassure them that there is a bias-free approach being taken. In Campbell and Finch (2004, p. 179) they discuss Tyler and Bies (1990) five factors that influence employees' perceptions of procedural fairness. These include:

- adequate consideration of the viewpoints of others
- consistency in the criteria on which decisions are based
- being bias free
- the provision of timely feedback and
- effectively communicating the basis for decisions.

This does not appear to be rocket science, yet if these five factors are not addressed they can be a major cause of disruption during facilities change processes. Quirke (1995) identified the need to ensure the 'why' is communicated to anyone involved in a change process, yet the justification for the change is often not communicated. Anecdotally, facilities are reasonably good at communicating with end users in relation to what is going to happen and when; but the 'why' is often forgotten. The issue of engaging the end user from the start of the process is also often forgotten: this can be as simple as an invitation to come and discuss the current issues, what could be improved, what would staff like to see? From a recent change experience within the author's own organisation, the use of comments boxes, after an initial announcement that there needs to be some form of change, can allow for voices to be heard. Alternatively if a major change is proposed, Quirke (1995) would argue that the medium that is used to communicate with the staff can have a huge impact.

8.2   THE PROJECT TEAM AND PREPARING THE STAGE

Pritchard (2007) discusses six elements of effective implementation in relation to learning events, but his ideas are just as key to working with colleagues/end users in relation to implementing a project.

1. 'Initial contact stage' — in relation to FM and change, this needs to be not only the development of a cohesive team to deliver the strategic objectives of the business but also the initial communication to end users about the change; this communication should take place early in the project, especially as people within the project team may not necessarily be in the FM discipline, but from the wider business. These people can also be considered as champions of the project to other end users.
2. 'Main focus/initiative', i.e. linking the ideas to reality. For instance, the reality of the move is clear and members of the project team understand their roles and the way forward and what is expected of them. Also, the end users begin to understand the wider picture and the 'selling of the ideas' has begun to happen.
3. The 'make it happen' stage reminds us of the momentum needed to keep the project moving: we should be enthusing the team and the end users to engage with the project but also ensuring an openness to listen to suggestions and feedback, remembering true communication allows for a two-way flow.
4. The 'review stage' equates to the results phase, for example as a project is rolled out there is an opportunity to engage with one of the initial groups impacted by it, identifying them as champions. In so doing, they are able to convey how it has enabled them to improve working practice.
5. Pritchard's need for 'communication' runs throughout his six elements. The next section will further focus on this.
6. The final phase is the 'measurement' phase, which is not confined to measuring the effectiveness of the project (time/cost/quality) but is also concerned with reviewing how the team worked together and any issues that arose? How effective were the communication methods used? Are there any lessons learned from the people elements of a project to move forward?

Tannenbaum and Schmidt (1958) created the *Continuum of Leadership Behaviour*, which identified the amount of authority linked to the manager and how much freedom was delegated to subordinates in decision making. The continuum is outlined below and includes ideas for FM to utilise during the process of move management. This concept focuses on a continuum with 1 being autocratic management and 7 being the most democratic.

1. Manager makes decision and announces it. The FM department has been charged with the move, and this is announced to the end users. This sits at the far end of the autocratic process — a decision has been made with no consultation of users and they are informed of the decision.
2. Manager 'sells' decision. The FM department need to engage the end users to buy into the move and therefore the advantages of the move need to be communicated to the end users. A decision has been made but rather than just telling the end users about the move, there is a deeper engagement with them as there is recognition that there may be reluctance or resistance to the change.
3. Manager presents ideas and invites questions. The FM department has decided what needs to be done but, rather than present a 'fait accompli', decides to share the information with the end users and invites comments, encouraging a two-way communication process, allowing a deeper understanding of any issues.
4. Manager presents a tentative decision subject to change. The FM department has a tentative decision on what needs to be achieved to meet its goals, is prepared to present

the proposed solution to the end users to hear their views, but reserves the right to still make the final decision.

5. Manager presents the problem, gets suggestions, and then makes the decision. Up to this point the FM department have come to the end users with a solution but using this approach the users are given an opportunity to explore possible solutions. The FM department still make the final decision.

6. The manager defines the limits and requests the group to make a decision. At this point the FM department passes responsibility to the group (which they may still be part of), defines the issues and the parameters but relies on the group to come to a decision that meets the end goals.

7. The manager permits the group to make decisions within prescribed limits. In the final stage, which is the most democratic, the FM department allows the end users to define the current issues and solutions and commits to working with the decision that is made.

As a discussion related to engaging with users, the latter end of the continuum may assist in reducing the resistance to change. It may not always be feasible to engage at the far end in complete democracy, but if you can speak to the end users and explain what is needed to achieve the end goals, be it cost saving, closing a building, etc., and ask them to make suggestions on how this could be done, then they have not only engaged with the process, they have ownership of it.

Following on from Pritchard's (2007) model and Tannenbaum and Schmidt's (1958) continuum in relation to the engagement of the project team and the end users, further consideration is given later in this chapter to the impact that communication can have on the implementation of any FM project. This is demonstrated by means of a case study in relation to the FM department of a private sector organisation, in which they analysed their communication following a large organisational workplace change and the impact that this had on the overall satisfaction of staff.

8.3   ALTERNATIVE WORKPLACE STRATEGIES AND SPACE UTILISATION

Large office-based organisations can incur significant property costs from its offices, yet they are frequently under utilised (Fawcett and Chadwick, 2007). Generally facilities management (FM) departments have been responsible for reducing real estate costs within organisations — a process that usually occurs through cramming more people into a space rather than using it more efficiently (Robertson, 2000). However, organisations need to move away from the one desk per person paradigm. As Melvin (1992) argues: new thinking demands different physical space; but it's not a simple case of getting rid of the existing properties and starting again (Chilton and Baldry, 1997). Robertson (2000) suggests that work transformation is required — the way work is carried out and the environment in which it occurs should be challenged. Becker *et al.* (1994) recognised that any changes to workspace that were cost driven as opposed to business enhancement driven usually resulted in reduced satisfaction levels.

Alternative Workplace Strategies (AWS) describe changes to workplace design, on-site working strategies and off-site working strategies (Gilleard and Rees, 1998). Typical examples of these (in order) are flexible working schedules (i.e. using the office for more hours in the day than 9 to 5), hot desking[1], and telecommuting (a mix of home and office-based working). Implementation of AWS could save an organisation 30% of its office costs (Mawson, 2009). There are three main resources that require consideration in any AWS – people, technology and space (Robertson, 2000). These resources are interdependent so if the way office space is used changes, there may be effects on both the people and technology.

Ratcliffe (2009) suggests that although cost savings are a genuine consequence of a workplace strategy they should not be the motivation for implementing it. In FinanceCo's case, cost reduction was the primary driver. The use of alternative space arrangements does not come without its problems, as increased space utilisation brings with it a number of workspace issues (Chilton and Baldry, 1997), i.e. space for filing. It is important to consider the employee in all this, as the working environment can affect morale (Wates Interiors, 2006), job satisfaction (Carlopio, 1996) and attitudes (Lee, 2006). The implementation of an AWS will affect cultural norms and the working environment (Chilton and Baldry, 1997) which could reduce both morale and satisfaction.

8.4 COMMUNICATION

La Framboise *et al.* (2003) discussed the 'four stages of change' process, including; (1) discovery – what is happening – a critical phase in the communication plan; (2) denial – it won't actually happen; (3) resistance – a time of employee concern, a time when employees need personal communication to understand how this affects them personally; and finally (4) acceptance. By this stage the communications plan should have eased the change process, and employees should be feeling fully informed and fully aware of how the proposed change affects them. La Framboise *et al.* (2003) also recognised the need to allow the opportunity for staff to feedback ideas, but also to show how their contributions have been valued.

Communication is a two-way process: information can be transmitted, but communication has to be shared (Quirke, 1995). Information is one-dimensional, i.e. newsletters or intranet messages (Van Vuuren and Elving, 2008); however, it lacks the engagement (Nutt, 1999) that when used appropriately creates a mutual understanding between parties (Elving, 2005). Holm (2006) suggests that communication is the process by which individuals share meaning through a transactional process between two or more parties.

8.5 CHANGE MANAGEMENT THEORY

Organisations typically suffer from a 'paradox of culture' (Lakomski, 2001) as they are constantly balancing the forces for change with the need for stability; this stalemate becomes broken when the force for change is greater than the resistance to it, causing a shift in the organisation's equilibrium (Lewin, 1951). By 'unfreezing' the existing behaviours and patterns, 'moving' them to the desired state and then 'refreezing' them it allows an organisation to successfully implement a change.

---

[1] Hot desking relates to the practice of providing a pool of equipped desks which are occupied as required instead of giving employees their own desk. Definition adapted from http://www.businessdictionary.com/definition/hot-desking.html.

Any change to the way space is used within an organisation will create an amount of uncertainty and fear amongst employees (Senior and Fleming, 2006). There can be a significant resistance to change as the way in which people operate is shifted in a direction that the existing culture may be uncomfortable with. Mawson (2009) describes this as ungluing the 'attitude glue' — achieved by removing people's 'psychological safety net' (Schein, 1987). By revising the traditional working method employees are opened up to the unanticipated consequences this change may bring (Nevis, 1987).

This feeling is very common in change programmes, where the initial shock is followed by denial and anger (Hopson and Adams, 1976). It is not until recognition of the inevitability of the change and the reasons for it are accepted, that the individual will be liberated from their past. This is the process of moving, where the new culture will evolve to a point where it accepts the new vision (La Framboise *et al.*, 2003).

When dealing with cultural change it is important to ensure there is a suitable balance between change and stability (Trice and Beyer, 1993), otherwise the level of resistance will restrict change. The best time for cultural change is when favourable circumstances exist, such as poor company performance or competitive markets (Trice and Beyer, 1993).

8.6   COMMUNICATION IN CHANGE MANAGEMENT

Communication is one of the most important aspects of change in the workplace (Lewis and Seibold, 1998, as cited in Allen *et al.*, 2007). Quirke (1995) takes this further by suggesting that communication has *the* pivotal role in managing change. Organisations often do not realise that without effective employee communication, change is impossible (Barrett, 2002). Unfortunately there is no universal approach to effective organisational change communication; it depends on the size, culture, style, stability and available resources (Daly *et al.*, 2003). To be effective, communication should be regular, timely, honest, clear, interactive and easy to understand with the opportunity for two-way communication otherwise the change programme may fail (Smith, 2006). There is a need to ensure real communication and not just a delivery of information.

Meaningful communication informs and educates employees at all levels and motivates them to support the strategy (Barrett, 2002). This is important as positive attitudes to change are vital in successful change programmes (Kotter, 1996), as resistance to change is one of the most significant barriers to overcome. Meaningful communication requires a degree of 'cognitive organisational reorientation' (Van Vuuren and Elving, 2008), i.e. comprehension and appreciation of the proposed change. There needs to be an understanding of the benefits of, for example, the new workplace and the part individuals need to play in it.

Burnes (1992), as cited in Kitchen and Daly (2002), states that communication is a way of avoiding uncertainty that change can provoke. This is important as high levels of uncertainty will negatively affect readiness for change (Elving, 2005). Resistance to change occurs when there is a lack of information or perceived benefits. People like to be in control of their destiny and outside change is a threat to this control which prompts resistance (Proctor and Doukakis, 2003).

Some organisations do not apply the same energy to communications as they would to the financial and operational aspects of change (Barrett, 2002). This is unfortunate as failure to engage with employees affected by change leads to a reduced chance of success (Ratcliffe, 2009), and increased hostility to the change. Employees can only be effective in the organisation if they are fully informed (Kitchen and Daly, 2002), particularly on the objectives of the change (Young and Post, 1993).

Elving (2005) proposes a communications model that displays the impact of communication on uncertainty and readiness for change. When the communication is meaningful and describes the reason for the change there is greater readiness, whereas if the information is vague it is more likely to create uncertainty. This is why personally relevant information is better than general information (Klein, 1996; Goodman *et al.*, 1996) when communicating for change. During the change process the communication needs of those affected by the change needs to be considered.

Messages regarding change are frequently communicated from senior level to employees by a cascade through what Klein (1996) calls the 'line hierarchy'. There are advantages in this process, as messages from those in power carry greater significance (Young and Post, 1993). But this method is not without its flaws. When information is channelled down the hierarchical pyramid the context often becomes lost so the people further down the organisation make less sense of what they receive (Quirke, 1995). The information received tends to become more about 'how the change will happen' rather than 'why it is happening'. Where studied this would appear to be true for facilities management teams and their delivery of changes to the working environment (La Framboise *et al.*, 2003).

Information is a commodity to be brokered and a resource to be guarded. Often this information is (at middle management level) withheld, changed, manipulated or delivered too late (Witherspoon and Wohlert, 1996). Quirke (1995) describes this as a 'refractive layer', whereas Waller and Polonsky (1998) depict these individuals as 'influencers' because their views manipulate the communication. The influencers create 'noise' in the communications process that distorts the intended message (Shannon and Weaver, 1949). Young and Post (1993) reported that communication carries more weight when it comes from someone in authority because the status of the messenger gives credibility to the message (Klein, 1996).

As Quirke (1995, 2000) suggests, communication is vital to ensure the individual is aware of how the change impacts and affects them. There is a need to be open to two-way communication, not just explaining how the new change will work in practice but also in listening to feedback and being open to challenges.

8.7   COMMUNICATION METHODS/MEDIUMS

The success of any change effort relies on how the message was communicated to the targets of the change (Witherspoon and Wohlert, 1996). Research by Allen *et al.* (2007) found that employees who indicated they received quality communication about the change demonstrated a more positive attitude towards it.

It is also important to acknowledge that the provision of information may not be sufficient to reduce employee uncertainty (Allen *et al.*, 2007). It is important that feedback (i.e. two-way information flow) exists rather than just dictatorial statements being made. It is important that management listens and takes time to communicate (Smith, 2006).

The facilities management team need to ensure that the correct medium is used in relation to the message: for example, if the change is major and there needs to be some form of discussion then a rich medium, such as telephone or face to face, should be used; however, if it is a minor change with little affect on the staff, then a lean medium could be used, such as posters or emails, etc. (Quirke, 1995).

8.8   CASE STUDY

FinanceCo is a 'blue chip' organisation that recognised in the recent economic climate that there was a need to reduce overheads and in particular facilities expenditure, as this is the second largest cost to a business after personnel (Williams, 1996). In an attempt to reduce and rationalise their property portfolio, FinanceCo carried out an alternative workplace trial in March 2008, based on sharing desks in team zones, rather than individuals having their own desks. Desks were allocated at the ratio of five desks per six full time employees (FTE), with an additional small number of shared touchdown points that anyone could use. After the initial trial there was anecdotal evidence that employees felt a significant amount of resentment and viewed the change as penny pinching. At no stage was a review undertaken to ascertain whether the pilot was successful or not and there was also no evaluation of the communication used to inform staff of the change.

For such a considerable undertaking, sufficient time and resources would ideally have been required — not only to develop and implement the solution, but also to communicate it. Unfortunately, the timescales for delivery were driven by lease end dates, which meant that although considerable time and effort had been put into the design of the office space and how it would be used more efficiently, the communication of how this change would be implemented and its effect on employees was not undertaken. The research reported in this chapter focuses on the communication used in the change programme and its impact on the relative satisfaction of the respondents. This change had the effect of cultural transformation, moving from having your own desk to sharing space. Undertaking an evaluation on the way in which this change was communicated provides useful evidence for future organisational change initiatives.

The hypothesis for the research was that a lack of a communication strategy had led to reduced staff satisfaction and therefore has negatively impacted on the change outcome. The survey was submitted to 1600 FinanceCo staff in March 2009, with a response rate of 32%.

The questions asked within the survey focused on understanding whether there was any correlation between the satisfaction of the alternative workplace strategy and the level of communication received. The results are discussed further in the following section.

8.9   COMMUNICATION METHODS USED

The property and facilities management department made the initial communication of the proposed change in the case study. This took the form of workshops with directors and heads of departments involved in the change: this occurred two months prior to the change taking place. The purpose of the change was explained and a detailed explanation of how it was to be achieved was outlined. There was a lack of understanding on the 'why' the change had to happen in relation to reducing the property portfolio to reduce costs to the organisation. The workshop participants were then asked to disseminate this information to their teams. Although this was the predominant route for communication, unfortunately some end users became aware of the alternative workplace strategy via other means (i.e. office rumour).

A number of approaches were used to communicate the alternative workplace strategy. However, 17% of the survey respondents believed that they did not receive any communication before the workplace strategy was implemented in their office. Verbal communication tended to predominate, but this did not follow the accepted communication norm in

FinanceCo where initiatives or news items that had a material impact on its employees were initially advised by a verbal briefing followed by an article on the intranet. In this instance there was no intranet follow up.

The results showed that there was no 'best' method for communicating the concept to all employees. There was also a variation of preferred methods based on the location and culture of individuals. For example, IT professionals supporting the business preferred email contact and people that interacted with customers by telephone preferred verbal communication, arguing that communication can not be a one size fits all approach.

From the data it would appear that exactly 'who' delivers the alternative workplace strategy communication (i.e. how the survey participants found out about it) determined the acceptance of the communication vehicle as an adequate means to deliver it. It was more likely the recipient of the information would be more satisfied with the communication method if it came from their line manager rather than the FM team or office rumour. This supports the view of Klein (1996) who suggests that 'line hierarchy', i.e. passing the message down the command chain, is the most effective communications channel. It was also important to ensure there was a single message rather than a distorted version of facts (people's subjective opinions).

8.10   FEEDBACK

In relation to the opportunity for feedback on the suitability of the alternative workspace strategy, only 24% of respondents (113 out of 473 people) felt they were able to offer feedback once the message had been conveyed (Figure 8.1). Of those 113 able to give feedback, only 72 (64%) of respondents were satisfied that their opinion was taken seriously (Figure 8.2). Part of the reason for the low feedback opportunity was the communication methods used. Face to face communication gave people a greater chance to provide feedback in comparison to email. There was also an issue of a sense of 'fait accompli' from employees due to the lack of initial consultation, thus making them feel that feedback was almost pointless. In the author's opinion, if staff had been engaged in consultation in the

Opportunity to feedback

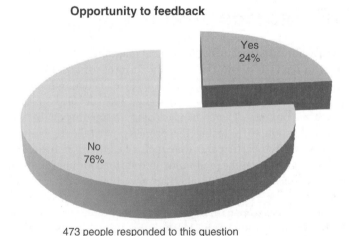

Yes
24%

No
76%

473 people responded to this question

Figure 8.1   Opportunity to feedback on suitability.

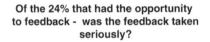

**Of the 24% that had the opportunity
to feedback - was the feedback taken
seriously?**

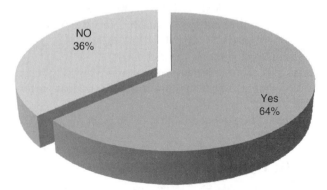

72 people resonded to this question

Figure 8.2   If the opportunity for feedback was given, was the feedback taken seriously?

initial stages it could have avoided the feeling of dissatisfaction in relation to the workplace strategy itself. Even though sometimes there is an absolute fait accompli, if people feel they have been given the chance to voice their opinions, then the resistance to change can be minimised, as discussed in the Tannenbaum and Schmidt (1958) continuum mentioned in Section 8.1.

8.11   SATISFACTION WITH METHOD USED

When the 'correct' communication method (a method that suited the individual) was used there was a greater level of satisfaction with the message in terms of content. Nelissen and van Selm (2008) suggested that responses to organisational change were influenced by management's communication of the reasons for and the consequences of the change. This concept also emerged from the data, where satisfaction with the way the reason for the change had been communicated was more significant than the frequency or amount. Personally relevant information was better than general information (Klein, 1996); thus explaining why reason and impact was more important to the survey respondents.

8.11.1   Communicating impact and reason for change

One member of staff quoted:

> The whole communication process from 'the estates and FM team' was very poor. There should have been consultation about its implementation — instead it was a 'tell' session and the needs of the business were not listened to or taken on board.

There was a significant variation in the quality of the communication. In instances of 'bad' communication the end user was either ignored, or the degree of detail was lacking, or the information arrived too late.

8.11.2 Suggested methods for improving communication

Participants were given the opportunity to propose how the communication of the alternative workplace strategy could have been improved. Other than responses that indicate the inadequacies of the strategy, the answers focused on four main themes:

1. communication should fully inform the end user
2. the way the message is communicated is important
3. the communication should provide the opportunity for feedback to take place and listen to it
4. the information should be provided in a timely manner.

8.12   SATISFACTION

The survey indicated that 43% of the respondents' satisfaction levels were 'worse' or 'much worse' after the change when compared to the original situation, with only 7% seeing any improvement. This was based on a retrospective opinion. Prior to the research there was an assumption that occupants that retained their desk would be more satisfied than those who had to work on a hot-desking basis, simply because there was less impact on those staff. However, the survey suggested that they were only marginally more satisfied and still displayed a degree of dissatisfaction with the organisation overall and the communication methods used.

8.13   COMMUNICATION OF CHANGE NOT APPROPRIATE OR EFFECTIVE

Verbal communication appeared to be the predominant method of informing employees, with 66% of respondents receiving the communication in this way; 25% by email; and 'other methods' represented the remaining 9%. Location, or rather the culture of the individuals working within a location, played an important role in identifying the most suitable communication method used. The data showed that the preferred approach was dependent on the culture and work styles of the individuals affected by change. Customer-facing staff preferred verbal communication, and back office IT support staff preferred email.

Initially this would be a sign of promising change management — the communication method used was matched to the staff involved. However, 17% of the survey sample believed they had not received any communication before the alternative workplace strategy was implemented in their office. This result would suggest a dysfunctional change communication strategy (based on the assertion of Quirke's, 1995) — because not everyone was aware of the change.

Although the alternative workplace strategy had executive level sponsorship, backing for the project was not communicated down the hierarchy (Klein, 1996). Instead, from a respondent's perspective, it originated from the property and facilities management department and therefore lacked the significance of messages derived from senior management (Young and Post, 1993). The research identified that respondents would have liked the message to have been 'cascaded from the top down'. However, using a cascade approach as the only method for communicating the alternative workplace strategy increased the opportunity for information to be withheld, changed, manipulated or delivered too late (Witherspoon and Wohlert, 1996). The directors and heads of departments were influencers of the communication flow. This gave rise to 'noise' in the form of their own opinions or objections — and might explain why some of the respondents were not aware of the change prior to its implementation. Alternatively, the influencers may not have had sufficient information at their disposal to effectively communicate the change as evidenced in some of the qualitative responses.

It may have been possible to prevent the adulteration and withholding of information if other communication methods had been used. Clearly using a single communication method did not satisfy the communication needs of FinanceCo employees. Evidence from the research, when respondents were asked which communication method they would have preferred, showed that they would rather there was a 'mixture of communication mediums' used.

In other FinanceCo change programmes, the intranet site and business unit emails have been used to convey the message — providing an effective short cut in the passage of information (Proctor and Doukakis, 2003). It was not clear why the intranet was not used to communicate this programme.

The research indicated that when the message was presented by appropriate people in a preferred manner it was more likely that the respondent would be satisfied with the content. Central to this concept is the importance of feedback — two-way information flow between sender and receiver. Face to face communication provides a greater opportunity for feedback (Quirke, 2000) and this is shown in the data: in particular, when the local manager was involved. Unfortunately, with the alternative workplace strategy, only a quarter of those surveyed had an opportunity to provide feedback.

Quirke (2000) suggests that it is important for those involved in change to understand the full picture — to know 'why' the change is happening rather than just dealing with 'how' it is happening. The alternative workplace strategy was promoted as a means of providing employees with flexibility in the workplace. However, comments from the research showed that there was no point dressing this up as anything other than cost saving.

Communication of the reason and impact of a change, in particular personally relevant information (as identified by Klein, 1996), was significant in influencing participants to accept the change. This research showed this was more important than either the frequency or amount of information supplied. Respondents were three times more likely to have received 'bad communication', (i.e. detail of the change was not adequate or the needs of the end user was ignored) when the impact or reason for the change was not explained. This may explain why the satisfaction of many respondents was low. The information they were looking for was often not presented and this did not allow them to personally engage with the change.

8.13.1　Case study: conclusion and recommendations

The purpose of this research was to test the hypothesis that the lack of a communication strategy leads to reduced staff satisfaction and therefore negatively impacted on the change outcome. The feedback from this research was that the respondent's requirements were not considered prior to implementation. From responses it was clear that the alternative workplace strategy itself was not a 'one size fits all' solution and comments from staff alluded to it being a flawed concept because it simply did not work for them. Therefore, given that there was no universal disapproval of the concept (half the respondents were no less satisfied) it would be inappropriate to say that the workplace strategy does not work. Instead what was apparent is that its implementation, even though it was founded on previous reports or trials, was imposed without sufficient consultation. The data consistently indicates that there was insufficient opportunity to provide feedback to those with influence. On the occasions where feedback was provided, the response was 'it's going to happen regardless — just accept it'. This would reinforce the findings from La Framboise et al.'s (2003) research that recognised the importance of having a robust communication plan in place.

Therefore, if the respondents had received enough information via a range of communication media, which had arrived in adequate time to understand the reason and impact of the change, together with the opportunity to participate in meaningful feedback, then the satisfaction levels of working with FinanceCo could well have been improved.

8.14   RECOMMENDATIONS

Sufficient time should be given to implementing any change, which should have a communications strategy at its heart. The full picture of why the change is necessary and what impact it will have on each individual should be conveyed using multiple means of communication in order to ensure everyone involved is aware and they receive the message without distortion. End users could also be engaged, as part of the project team, to ensure the business needs are being met within the proposed new workspace. This also creates 'champions' for the project within the business.

Verbal communication is important, particularly when delivered by line management', as not only is there a degree of trust in the content of the message but employees will have a greater opportunity to comment. True communication is a two-way process, and therefore listening skills are as important as information giving. Feedback must also be taken seriously and responded to.

Once the change programme has begun, evidence from the study suggests that change agents should not stop communicating. Staff needed to be informed of progress and given the opportunity to seek answers in order to justify the change. When the change has finished, evaluation of the project and the communication strategy should be undertaken. In turn, any recommendations from the evaluation, and especially the communication used, should be taken forward to the next change initiative.

REFERENCES

Allen, J., Jimmieson, N.L., Bordia, P. and Irmer, B.E. (2007). Uncertainty during organizational change: Managing perceptions through communication. *Journal of Change Management*, 7(2), 187–210.

Barrett, D.J. (2002). Change communication: using strategic employee communication to facilitate major change. *Corporate Communications: An International Journal*, 7(4), 219–231.

Becker, F., Quinn, K.L., Rappaport, A.J. and Sins, W.R. (1994). *Implementing Innovative Workplaces. Organizational Implications of Different Strategies*. Ithaca, NY, Cornell International Workplace Studies Program.

Burnes, B. (1992). *Managing Change*, 2nd edn., Pitman Publishing, London.

Campbell, L. and Finch, E. (2004). Customer satisfaction and organisational justice. *Facilities*, 22(7/8), 178–189.

Carlopio, J.R. (1996). Construct validity of a physical work environment satisfaction questionnaire. *Journal of Occupational Health Psychology*, 1(3), 330–344.

Chilton, J.J. and Baldry, D. (1997). The effects of integrated workplace strategies on commercial office space. *Facilities*, 15(7/8), 187–194.

Daly, F., Teague, P. and Kitchen, P. (2003). Exploring the role of internal communication during organisational change. *Corporate Communications: An International Journal*, 8(3), 153–162.

Donald, I. (1994). Management and change in office environments. *Journal of Environmental Psychology*, 14 (1), 21–30.

Elving, W.J.L. (2005). The role of communication in organisational change. *Corporate Communications: An International Journal*, 10(2), 129–138.

Fawcett, W. and Chadwick, A. (2007). Space-time management and office floorspace demand: Applied experience and mathematical simulations. *Journal of Corporate Real Estate*, 9(1), 5–27.

Folger, R. and Cropanzano, R. (1998). *Organizational Justice and Human Resource Management*. Sage Publications, Thousand Oaks, CA.

Gilleard, J.D. and Rees, D.R. (1998). Alternative workplace strategies in Hong Kong. *Facilities*, 16(5/6), 133–137.

Goodman, M.B., Holihan, V.C. and Willis, K.E. (1996). Communication and change: Effective communication is personal, global and continuous. *Journal of Communication Management*, 1(2), 115–133.

Holm, F. (2006). Communication processes in critical systems: dialogues concerning communications. *Marketing Intelligence & Planning*, 24(5), 493–504.

Homans, G.C. (1961). *Social Behavior: Its Elementary Forms*. Harcourt, New York, NY.

Hopson, B. and Adams, J.D. (1976). Towards an Understanding of Transition: Defining Some Boundaries of Transition Dynamics. In: Adams, J.D., Hayes, J., Hopson, B. (eds.) *Transition: Understanding and Managing Personal Change*. Martin Robertson.

Kitchen, P.J. and Daly, F. (2002). Internal communication during change management. *Corporate Communications: An International Journal*, 7(1), 46–53.

Klein, S.M. (1996). A management communication strategy for change. *Journal of Organizational Change Management*, 9(2), 32–46.

Kotter, J.P. (1996). *Leading Change*. Harvard Business School Press.

La Framboise, D., Nelson, R.L. and Schmaltz, J. (2003). Managing resistance to change in workplace accommodation projects. *Journal of Facilities Management*, 1(4), 306–321.

Lakomski, G. (2001). Organisational change, leadership and learning: Culture as a cognitive process. *The International Journal of Educational Management*, 15(2), 68–77.

Lee, S.Y. (2006). Expectations of employees towards the workspace and environmental satisfaction. *Facilities*, 24(9/10), 343–353.

Lewin, K. (1951). *Field Theory in Social Science*. Harpers.

Lewis, L. and Seibold, D. (1998). Reconceptualizing organizational change implementation as a communication problem: a review of literature and research agenda. *Communication Yearbook*, 21, 93–151.

Mawson, A. (2009). Sticking point. *FM World*, 6(8), 24–26.

Melvin, J. (1992). Offices for the 1990s. *Facilities*, 10(11), 16–19.

Nelissen, P. and van Selm, M. (2008). Surviving organisational change: how management communication helps balance mixed feelings. *Corporate Communications: An International Journal*, 13(3), 306–318.

Nevis, E.C. (1987). *Organizational Consulting: A Gestalt Approach*. Gardner Press.

Nutt, P.C. (1999). Surprising but true: Half the decisions in organisations fail. *Academy of Management Executive*, 13(4), 75–90.

Price, I. and Fortune, J. (2008). Open plan and academe: pre- and post-hoc conversations. In Finch, E. (ed.), *Proceedings of the CiB W70 Conference*. Edinburgh Herriot Watt University.

Pritchard, N. (2007). Efficient and effective implementation of people related projects. *Industrial and Commercial Training*, 39(4), 218–221.

Proctor, T. and Doukakis, I. (2003). Change management: The role of internal communication and employee development. *Corporate Communications: An International Journal*, 8(4), 268–277.

Quirke, B. (1995). *Communicating Change*. McGraw-Hill.

Quirke, B. (2000). *Making the Connections: Using Internal Communication to Turn Strategy into Action*. Gower.

Ratcliffe, P. (2009). Prepare and prosper. *FM World*, 6(8), 19.

Robertson, K. (2000). Work transformation: Integrating people, space and technology. *Facilities*, 18 (10/11/12), 376–382.

Schein, E. (1987). *Process Consultation II.* Addison-Wesley, Reading.

Senior, B. and Fleming, J. (2006). *Organizational Change,* 3rd edn., Prentice-Hall.

Shannon, C.F. and Weaver, W. (1949). *The Mathematical Theory of Communication.* The University of Illinois Press.

Smith, I. (2006). Continuing professional development and workplace learning — 15. *Library Management,* 27(4/5), 300—306.

Tannenbaum, R. and Schmidt, W.H. (1958). How to choose a leadership pattern. *Harvard Business Review,* 36(2), 95—101.

Trice, H.M. and Beyer, J.M. (1993). *The Cultures of Work Organisations.* Prentice-Hall.

Tyler, T.R. and Bies, R.J. (1990). Beyond formal procedures: The interpersonal context of procedural justice, in Carroll, J.S. (ed.), *Applied Social Psychology in Business Settings,* Erlbaum, Hillsdale, NJ, pp. 77—98.

Van Vuuren, M. and Elving, W.J.L. (2008). Communication, sensemaking and change as a chord of three strands. *Corporate Communications: An International Journal,* 13(3), 349—359.

Waller, D.S. and Polonsky, M.J. (1998). Multiple senders and receivers: A business communication model. *Corporate Communications: An International Journal,* 3(3), 83—91.

Wates Interiors (2006). *My office depresses me!* Available online at: http://www.onrec.com/content2/news.asp?ID=11256. Accessed 31st May 2011.

Williams, B (1996). Cost-effective facilities management: A practical approach. *Facilities,* 14(5/6), 26—38.

Witherspoon, P.D. and Wohlert, K.L. (1996). An approach to developing communication strategies for enhancing organizational diversity. *The Journal of Business Communication,* 13(4), 375—387.

Young, M. and Post, J.E. (1993). Managing to communicate, communicating to manage: How leading companies communicate with employees. *Organizational Dynamics,* 22(1), 31—43.

9 User Empowerment in Workspace Change

Jacqueline C. Vischer

CHAPTER OVERVIEW

Facilities management is often seen as the instrument of change in organisations. However, this relatively new profession is itself undergoing a revolution, with the emergence of participative involvement of building users in change management decisions. This chapter outlines a user-centric approach to building operations and management in which the identifying, managing and sharing of building-related information are key. As more information is shared with users, so employees have the potential to become more empowered to make decisions about their own workspace. The chapter describes how users resist change, and how the negative energy of resistance can be transformed into a positive force. Empowering building occupants leads to positive attitudes that generate a constructive and creative space planning process. This helps facilities managers achieve the objective of providing workspace that both meets users' needs and is a tool for getting work done.

Keywords: Workspace; Space planning; Users' needs; Empowerment; Transformation.

9.1   THE 'SCIENCE' OF USER PARTICIPATION

Traditionally, one of the concerns that architects bring to building projects is *utilitas* — that is, the functionality of the space created. The other two, as any architecture student knows, are *venustas* or beauty and *firmitas* or structural soundness. As buildings have become more complex, and the process of designing and building them requires increasing numbers of specialists and experts, so those responsible for buildings — the designer to begin with, then the builder, then the manager owner or operator — have searched out a variety of ways of assuring *utilitas*, that is, that the built space supports the uses to which it is put.

   In more modern times, the 'building user' has become a key source of specialised information about building use and function. Sometimes called the 'end user' in recognition of the parallel with computer technology and software development, the people who use the building for its intended function have become one of the specialised knowledge groups whose input, feedback and opinions are increasingly taken into account by building industry professionals.

Facilities Change Management. Edited by Edward Finch.
© 2012 Blackwell Publishing Ltd. Published 2012 by Blackwell Publishing Ltd.

One of characteristics of today's building industry is the evolution of systematic and effective ways of involving users and engaging their knowledge in useful and constructive ways. An entire sub-speciality of social psychology has grown up over the last 40 or so years, based on gathering knowledge about how humans relate to — behave in, think about, perceive and assess — built space. Many of the activities of the discipline of environmental psychology are aimed at ways of applying such knowledge practically to the design, planning, construction and operation of buildings (Gifford, 2006). Such activities include collecting feedback from occupants on how the built environment they occupy functions (post-occupancy evaluation), asking occupants to rate various environmental qualities and properties in terms of their own comfort and satisfaction (user surveys), and engaging future occupants in various strategies of building decision making, both in the design phase (programming or brief-writing), during detailed design and post-construction (commissioning and facilities management) (Preiser and Vischer, 2004). Collectively, these endeavours might be labelled forms of 'user participation'.

As part of the evolution of these practices it must be noted that, for various reasons, engagement of users is not widespread in the industry. Such reasons, which may or may not be based on evidence, include:

- concerns about the extra time it might take to consult users,
- liability of owner and managers if information from users is collected and not used, and
- cost issues if users, when questioned for information, profit from the opportunity to demand a long list of features not included in the project budget.

In addition, some project teams have trouble identifying future users, and some projects have other priorities, such as building speculatively for maximum profit, so that user participation is a non-starter.

However, as the field of environmental psychology has evolved, so have techniques and strategies for engaging building users in ways that are both time- and cost-effective. This results in long-term cost savings because wrong decisions are corrected earlier rather than later and costly mistakes avoided. One example is the rising interest in and acceptance of evidence-based design (Hamilton and Watkins, 2009). Others focus on ways of accessing users' unique knowledge of what they will do in the spaces provided, and how they will do it. When such information is rapidly and accurately applied to building decision making, users themselves have an opportunity to participate, to have a say in the design of the environment they will occupy, and to take ownership of the space and its *utilitas*.

9.2 FACILITIES MANAGERS AND USER PARTICIPATION

Some of the earliest supporters of user participation in design were members of the design professions. Indeed, many buildings were not built without extensive user participation in decision-making — designing houses for example (Stevenson and Leaman, 2010). However, involving users on an ongoing basis is increasingly becoming an option and even a decision-making tool for facilities managers of commercial buildings — places where people work — particularly where change is anticipated or required.

While facility managers have expertise in building management and operations, many also find themselves managing building users. Most aspects of building operations have pronounced technical requirements; but when users are uncomfortable with their thermal comfort, or when a change in use or configuration is planned, or when simple logistics (such as access to parking or bathrooms, slow elevators or storage requirements) become issues for occupants, then facility managers find themselves involved in more than the technical operations of the building.

This is particularly noticeable when user groups decide to change their environment, usually by adding or subtracting workstations, or by reconfiguring to allow more space for meetings, or when organisational shifts require groups to merge or separate. Facility managers often find themselves not only managing change to the physical environment, but also at a social and organisational level. The process of changing the physical environment cannot proceed in an effective manner without engaging users in the change process, and in so doing, discovering that the workspace change process benefits from empowered users.

The building occupant is in some ways the facility manager's best resource. Proactive facility managers seek out feedback from occupants to help them maintain comfortable conditions inside the building, and also to ensure that FM services meet occupants' requirements and expectations. In addition, proactive facility managers understand that while they are the experts on operating the building, the occupants are experts on the work they are doing and on the tasks they are required to complete in the environment the building provides. Increasingly, therefore, facility managers seek out more of a 'partnership' relationship with building users, in which occupants provide useful and constructive feedback to facility operators that help the latter not only provide a more comfortable and supportive environment, but also devise workspace changes that result in a more effective work environment for occupants, making them more effective and productive in their work (Vischer, 1996).

This partnership between users and managers needs to be carefully designed and structured to ensure that it is beneficial to both sides. Inviting user participation in environmental decision making can, if not well managed, lead to a prolonged and inconclusive process, to raised expectations that are not met, and to other negative outcomes that have undesirable consequences. On the other hand, a clear and explicit process, in which objectives are known, feedback is sought in a specific form on specific issues, and information is exchanged openly on both sides, can only improve decisions and increase the quality of workspace outcomes.

9.3   THE NEW WORKSPACE OPPORTUNITY

The New Workspace Opportunity (NWO) is offered every time workspace is changed, even if the scale of physical change is modest (Vischer, 2005). Every change to the environment offers users an opportunity to rethink their processes, their tools and the organisational structure so as to make themselves more effective at what they do. Facility managers need to be aware of the NWO, to help occupants take advantage of it and to provide the tools and skills needed to make it effective. Some of the tools employed in such a process include:

- a managed survey of occupants with a specific set of reliable questions whose answers can be applied to design decision-making,

- focus groups in which users are encouraged to identify environmental elements they find both supportive and not supportive to their work, and
- ideas sessions — oral, written or on-line — where members of the target user group are invited to suggest solutions to problems identified by designers and planners as part of the process.

These and other similar tools are ways of eliciting effective and constructive suggestions from users without encouraging users to generate their 'wish list' and unduly raising their expectations (Vischer, 2005). Typically, employees at work do not seek out additional information about their workspace, the building or how it operates. FM controls this information, and can disseminate it if needed during workspace change. When renovations or moves to new premises are being planned, users benefit from information that otherwise they invent themselves. Such fabricated facts can be used to develop negative scenarios about the new space, based on occupants' fear of change.

Unfortunately, pressures to reduce the time and money spent on new workspace mean that participants do not always have the resources to systematically analyse and review opportunities for solving problems and improving processes before they move. Not all companies are equipped to analyse their existing business processes with a view to improving them; in fact not all organisations are rational enough to be motivated to do so (Mintzberg, 1994). Key workspace decisions often end up being made on the basis of lowest construction (not operations) cost, and time constraints. However, even a conventional design process anticipates a payback to the organisation in terms of more effective work performance, lower staff turnover and higher staff morale — all of which can and should be considered when making decisions about new space.

As in all change management processes in organisations, managing the acquisition, dissemination and flow of information is critical. The planning and design process for new workspace has specific information requirements at different stages. Building occupants and decision makers also need to be informed as the process advances. The information that is acquired and applied needs to be accurate and appropriate — and therefore from reliable sources — as well as up-to-date and unambiguous. Very often the information available is neither adequate nor reliable, so hearsay and gossip risk replacing the facts, and decisions get made based on fear and suspicion.

The facility manager has a strong voice in how information is gathered, who receives it and in what form, and how it is applied to planning and design. In renovations to existing buildings, FM staff provide key information about building maintenance and operations, about energy management objectives and sustainability, about floor loadings and bearing walls, about wiring and cables — power and data — (both vertically in a multi-storey building, and horizontally distributed on each floor) — and they specify cleaning and security requirements. FM staff are often responsible for managing access to users for the design team and others who need information about how the organisation operates and what — if any — business process changes are planned. Thus FM can and often do control the degree to which the organisation takes advantage of the New Workspace Opportunity when designing new workspace.

FM professionals in some organisations have developed a relationship with building users that goes beyond simply providing services related to the building and policing access and uses of space and facilities. FM personnel who have been entrusted by senior managers to occupy a strategic role in terms of organisational planning and future change maintain a more intimate partnership with building users. This enables them to understand the varying needs and requirements of different business unit managers and to respond to managers' requests in ways that help them solve problems and improve business processes. In many cases they do this by implementing systematic feedback from building users regarding their occupancy experiences, requirements and comfort.

Feedback from building users often takes the form of questionnaire surveys administered to occupants. The content of such surveys can vary from simple opinion polling: 'Was the FM service you received excellent, adequate or unacceptable?' to more complex questions concerning occupants' feelings of well-being, satisfaction and comfort, or lack thereof, as a function of building conditions. Occupant surveys that are performed on a regular basis provide FM staff with a diagnostic tool to assess what is working well and what is problematic for different groups performing different types of task in different parts of the building. When the questionnaire instrument is well designed and reliable, the information obtained from users provides a diagnostic profile of how effectively various building features are performing in terms of users' task requirements. Workspace conditions that are problematic — for example, indoor air quality problems in some areas, or glare from overhead lights in others, or heat gain and solar glare from windows — then become priorities for change in the workspace redesign process. Those building features that show up on a regular basis as causing discomfort or fatigue clearly affect occupants' performance and therefore the productivity of the group. Investment in workspace redesign solves many of these problems and improves occupant comfort levels, thereby assuring a legitimate objective for any workspace change project (Vischer, 2007).

In addition, facility managers may also acquire and disseminate feedback from occupants that has implications beyond the functional effect of specific building conditions and affects business processes. Almost any employee at any level has a few ideas about improvements that would enable them to work more effectively. Sometimes these ideas are limited to working physically closer to another individual or group, or, indeed, being more separated; sometimes these are more far reaching, and pertain to the location of and access to equipment, to ways of communicating and sharing information, and even to organisational restructuring. A common example is the need for more abundant and accessible places to meet and work together with one or a few colleagues on an as-needed basis. Conventional office space has often failed to meet this need, and the lack of collaborative workspace slows down decision-making and limits information exchange — priorities in the modern business environment.

The planning and design team may survey or interview building occupants to gain this type of information, but time constraints often prevent them from developing ideas with employees about ways of improving the functional support they receive from their workspace. The FM team who has already acquired this information and understood its importance to the overall health of the organisation is better placed to share it and ensure that the consultants prioritise it in their design decision making.

In maintaining an intimate awareness of their constituents' needs and aspirations, facilities managers are in a position to manage the all-important exchange of relevant information. Design decisions that have already been taken, such as a more open concept workspace, a building or site selected for the new space, or a change or upgrade to more powerful IT tools, need to be shared with building users. When they are not informed, such 'information' is fabricated by employees and moves quickly in the form of rumours throughout the organisation, convincing employees that they will have to relocate to another town or suburb, that there will be no enclosed offices, or that their parking privileges are to be removed, regardless of whether or not any of these are likely to happen.

This is not to say that sharing all information and all design decisions entirely eliminates employees' concerns about the future of their workspace. In spite of the rational need to identify workspace features that support work, people also have emotional attachments to the space they know, whatever its limitations, and an urge to resist any changes that risk altering the status quo.

9.4 PRINCIPLES OF WORKSPACE TRANSFORMATION

In all situations of workspace change, whether these are minor changes involving rearranging or replacing workstations or major changes involving completely renovating a floor or moving to new premises, employees are likely to express resistance to that which is new and unknown. While studies of organisational change largely focus on contextual conditions as drivers of change, research has also examined the pressure of organisational change on managers and their relative abilities to manage the stress this creates (Judge *et al.*, 1999). The study reports on seven personality traits that affect how managers respond to organisational change. Results show that of these, 'positive self-concept', made up of locus of control, generalised self-efficacy, self-esteem and positive affectivity, and 'risk tolerance', made up of openness to experience, tolerance of ambiguity and risk aversion, are strong dispositional predictors of managers' ability of cope well with change.

Similar arguments may be advanced to explain how well building users cope with workspace change. As with organisational change, workspace change generates resistance based on fear of the unknown — an effect especially pronounced where users are insufficiently informed. However, it can be argued that fear of workspace change is expressed in territorial terms, motivating users to defend territory and generating resistance to possible territorial loss and all that it means. While there are certainly personality differences that explain why some occupants may react more defiantly than others, there are also basic stages in coming to terms with workspace change that are common to all office occupants, and have some parallel to Kubler-Ross's process of accepting dying (Kübler-Ross, 1973). Typically, the stages of resisting and then coming to accept workspace change include:

- fear of loss
- mistrust of superiors
- identifying a champion
- too much change
- engaging in the process
- managing costs
- learning new processes (Vischer, 2005).

FM personnel can learn to anticipate these stages and to manage and distribute information within the context of the strategic user involvement process. The advantage of a proactive approach is fewer negative impacts arising from moving workers into an environment they did not anticipate and do not understand. Such negative impacts include importing previous processes and ways of working that no longer fit, anticipating a degree of comfort or luxury that is not provided, or simply not being able to find key contacts and resources that they need for their work.

Just as there are predictable stages to users' resistance to workspace change, there are opportunities throughout the design and planning process to take advantage of the opportunities these offer for constructive and useful feedback from users. Each stage of resistance or conflict offers a chance to empower users and give them a stake in the outcome. Typical stages of managing user resistance in a workspace change process are described below, and suggestions are offered for constructive FM responses that manage conflict and improve process outcomes.

9.4.1 Transformation as imperative

Companies cannot afford not to take advantage of the opportunity for organisational change and improvement that new workspace provides. The process of transforming workspace needs to articulate objective measurable improvements to the efficiency and effectiveness of organisational and work processes. All reconfiguration offers an opportunity for improvement, not just to space but at all levels of organisational functioning.

In acknowledging that space change is a tool for organisational change, managers need to commit to a rational analysis of how much and what sort of change they favour, but this is not easy to bring about. Taking a rational approach to planning workspace change can be frustrating in view of the emotional attachment people have to space. For example, employees at Boston Financial had their own offices, but the overall space configuration meant that very few of them had windows and most of the rooms had limited space. People sat in virtual closets, alone and cut off from colleagues, they were often too warm and were subjected to vibration from the air handling system located in the building's core. But when a major renovation was planned to open up the floor, give everyone a daylit space with access to windows and move them away from the building core and into a light and airy open space with adequate ventilation, employees rebelled on the basis that they would be accommodated in workstations and not in private offices. This is not an unfamiliar scenario. Rational thinking would assert that better light and air, not to mention ergonomic furniture, would help people work. Emotional attachment asserts that any enclosed space that can be viewed as 'private' is a marker of personal territory and status in the organisation and therefore should not be given up (Duffy *et al.*, 1998).

On the other hand, a complete vision of the future is not necessary for the process to be effective. Answers to smaller and more manageable questions about who we are and where we are going — and where we want to live — may be sufficient in minor move or change situations. Nevertheless, whether they see it at the beginning, middle or end of the process, decision makers ultimately recognise that planning workspace is a powerful tool to be used to achieve a variety of ends.

As mentioned above, feedback from users is a critical ingredient of the New Workspace Opportunity; the need for user involvement increases with the degree of change envisaged. At the small end of the scale of change, feedback from occupants may be applied to improving thermal comfort or lighting, or making sure better amenities — a windowed coffee-room, a quiet room or 'serenity space' — are included. In more major projects, where workspace change is part of a cultural transformation, user involvement should increase and users empowered to participate in decisions that affect them. As indicated in Figure 9.1, the scale of user involvement should keep pace with the scale of opportunity. Minimally, employees are consulted; more ambitious consultation leads to involvement in decision making; and at the large end of the scale of workspace transformation, employees need some

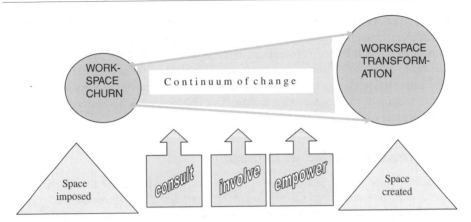

Figure 9.1 The new workspace opportunity. From Vischer, J.C. (2005). *Space Meets Status: Designing Workplace Performance* London, UK: Routledge/Taylor and Francis, Chapter 6.

control over decisions. Examples of the various tools and techniques for consultation, involvement and empowerment include surveys, interviews, focus groups and various forms of brainstorming and ideas sharing (Dewulf and Van Meel, 2003).

9.4.2 Play out the process

People at different levels and with different functions in the organisation react differently to the idea of workspace change, and to the idea of changing organisational culture. Stakeholders in the change process range from members of 'the C-suite' who may have a vision of the future of the organisation they want to implement, to middle managers and departmental leaders who have to make decisions about the space their staff will occupy while at the same time protecting their own status and territory. Employees are also stakeholders, and often stand to gain or lose most from new space. FM staff also have a point of view, often tied to practical considerations of cleaning and maintenance, as well as financial concerns as they often manage the budget for new construction as well as for facilities operation. Finally, consultants such as designers and space planners have opinions based on their specialised knowledge and experience.

Because each interest group has different priorities, a major challenge is to find ways of aligning stakeholders' priorities so that the project advances towards real change and does not revert to the 'tried and true' (see next principle). Not all stakeholders want change, or want the same amount of change, and some actively resist change. Moreover, their reasons for supporting or resisting change are likely to vary. As Figure 9.2 shows, sometimes senior managers favour change, whereas middle management and employees do not share the vision and are more inclined to protect what they have already. Sometimes the initiative for change comes from employees, who might or might not be supported by their managers, and who then have to convince senior management that change is desirable. In some projects the impetus for change is driven by consultants who gather enough wide-ranging information about the organisation to see where the opportunities for improvement lie. They might be supported by the FM people that hired them, but, in order to be effective, senior members of the organisation have to be convinced as well.

Stakeholders are aware that organisational culture and values are transmitted through workspace and building decisions (O'Mara, 2000). Both pro-change and anti-change

YES / NO	Senior management	Middle management	Employees	Facilities staff	Design consultants
Senior management					
Middle management					
Employees					
Facilities staff					
Design consultants					

Figure 9.2 Consensus matrix on transforming workspace. From Vischer, J.C. (2005). *Space Meets Status: Designing Workplace Performance* London, UK: Routledge/Taylor and Francis, Chapter 6.

proponents must understand that not all aspects of an organisation's culture need to be thrown out when change is envisioned: part of workspace transformation is to decide what to keep and what to change. Some companies have important cultural ties to their communities, and do not want to lose them. Others have traditional corporate rituals and symbols that give the corporation a human face to its employees. The recently built corporate headquarters of the Caisse de Dépôt et Placement in Montreal, Canada, has imprinted all the words that indicate its corporate values on the glass walls that soar above the six-storey atrium. The magnified letters are projected into the interior by the light outside, and serve as reminder to all employees and all visitors of the values the company stands for.

In the process of deciding what to keep and what to drop, the 'yes' groups and the 'no' groups have to negotiate, and the negotiating process has to be facilitated. New workspace means change, no matter how carefully stakeholders attempt to maintain the status quo, so an effort should be made to ensure that the changes that are implemented are advantageous to the organisation. The fact that all stakeholders have a different attitude towards and tolerance of change is an indicator of the importance of the process selected, and an indicator of the inevitability of conflict.

9.4.3 Embrace conflict

Conflicts in the change management process are typically about territory: territorial gains and losses, boundaries and claims. Changing territorial boundaries (and claims) inevitably causes conflict, but the energy released by conflict can be dynamic and constructive if it is recognised and harnessed by the change management process.

Embracing and resolving conflict allows the organisation's degree of tolerance for change to be identified. However, not all conflicts are resolved without some cost. Sometimes employees feel strongly enough about losing their private offices or their proximity to the window or their visitor's chair that they leave their jobs. Sometimes conflict is expressed in the form of threats: signed petitions, union action, official (and unofficial) complaints. One Canadian provincial government agency filed a grievance against management for moving staff out of 2.4 m by 2.4 m workstations with 1.5 m partitions into 1.8 m by 2.4 m work-stations with 1.2 m partitions. They claimed they would no longer be able to do their work, as the worksurfaces would be too small and the noise level would be too high. If senior managers ignore such concerns, employees will continue to resist workspace change; if they give in to such actions, the new workspace will be a compromise that will leave everyone unhappy. The conflict needs to be managed.

Tactics to get people talking about space and making their values explicit include presenting extreme scenarios ('let's take away all partitions around cubicles'), issuing status challenges ('let's put all the senior executives in open workstations') and framing futuristic scenarios (wireless throughout the building, remote work, shared desks, mobile offices, new intranet portals). These tactics are an effective way of getting stakeholders to engage in debate that makes their values explicit, but are more easily employed by outside consultants with experience in using them, rather than by internal FM personnel that may fear making enemies.

Conflict intimidates many people and in many organisational cultures overt conflict is considered wrong or shameful — somehow it is seen as evidence of failure. Conflicts that focus on space are a less threatening and less problematic way to express value differences, making people's differing viewpoints more explicit and more manageable than when abstract organisational values are at issue.

9.4.4 Avoid the default

The language that clients and users use ('demand'), and the language that building professionals and designers use ('supply') are not the same (Blyth and Worthington, 2001). One effect of the difference is that at certain critical decision points along the way, clients and users do not have a response to the questions they are being asked and cannot provide needed feedback to the design team about space use. They cannot easily envision future space: what a new place will look like, how it will affect them, whether or not they will be able to work comfortably. This knowledge vacuum creates pressure to go back to what is already known — the default position. This pressure exists throughout the workspace change process.

Requiring participants to be explicit about the goals and business objectives of workspace change is useful, especially at the beginning of the process. But it is not enough — these must be documented and accepted by all stakeholders. If not, they will remember later on important priorities they did not think of at the beginning, and they will forget the important priorities that they set at the beginning. Writing decisions down, making lists of goals and objectives, drafting design principles to guide later stage decision making — all these are tactics that help move the process forward and help avoid an automatic reversion to 'tried and true' solutions, which are, for the most part, repeat versions of what they already have.

Evidence of the pressure to default to the tried and true occurs when decision makers use cost and the possibility of cost overruns to avoid an innovative solution. Whether or not participants' new ideas are really 'pushing up costs', invoking this argument can effectively apply the brakes to real change. The whole question of whether or not innovative workspace is more costly than conventional offices and conference rooms is more complex than the easily understood but incomplete conventional real estate formulas such as cost-per-square foot and net useable-to-gross floor area. If workspace innovation, like new technology and business processes, is predicated on improving worker performance and making tasks easier and quicker to perform, then 'costs' are really 'investments' and a return should be expected, rather than a loss.

9.4.5   Not a zero-sum game

Fear of loss is an area of territorial conflict. Workspace occupants often tend to think more about what they are going to lose than what they might gain from new workspace. Anticipating and managing this expectation is an important part of managing workspace change. Building occupants tend to fear that organisational gains — such as reducing occupancy costs, opening up workspace and facilitating communication — are more likely to be gains to the company and not to themselves. They may experience these 'gains' as losses: smaller workspaces, fewer partitions and being more closely observed at work. Managing the workspace change process means taking steps to help users understand and believe that ultimately there are gains to both sides.

The key to creating a win—win situation is the dissemination of information. If gains in functional comfort and productivity, as well as cost savings, are the objectives of the innovative workspace project, then all stakeholders must be informed. It can take months after moving before people who fought against smaller workstations, lower partitions and clustered lay-outs find out that they like being close to co-workers, hearing others talk without having to call meetings, and being able to access other spaces when they need to concentrate or meet privately. Hypertherm Inc. is one of many organisations that had so outgrown its workspace that people were working in workstations squeezed into hallways, had lost virtually all their meeting-rooms and informal space, placing tables in hallways to have a place to meet, and were constantly too warm and experiencing poor indoor air quality because of the overly dense occupancy. Yet when a new building was being planned, several groups complained that it would be 'too open' and would prevent them from concentrating — not because this was likely to be true, but because they were moving into an unknown environment and they feared territorial loss.

The experience of loss is part of the change process — people resist loss, they fight not to have loss, they feel grief when loss occurs. Gains are not automatic — some teaching and preparation is needed so that people learn to function in new ways and take advantage of the interesting new possibilities their new space presents. While the threat of territorial loss automatically accompanies all workspace change initiatives, the possibility of gains must be learned. The most powerful antidote to territorial loss is empowerment through information and involvement.

9.4.6   Empowerment is key

People affected by change need to have some involvement in order for change to be successful. Because people's territorial rules and boundaries are emotional, users need some control over decisions, and to be effectively involved they need to be informed. This is what is meant by environmental empowerment. Deciding which decisions they have a say in, at which stages in the process they are consulted, and how much control they have over the outcome are key issues in the design of the strategic user involvement process.

There is no one right way of empowering people to make decisions about their workspace. The concept of empowerment is based on user feedback and consultation opportunities designed to be effective within specific time and cost parameters and to respond to real project constraints. By being clear and explicit about how much control users have and when they may exercise it, facilities managers can perform the important act of actually giving away control for that part of the process. A wide range of styles and degrees of empowerment have been developed, drawing from a variety of participation techniques and selected to correspond to the scale of transformation opportunity (Sanoff, 1999).

A major advantage of involving users is that their professional and technical knowledge can inform design decisions. Workers themselves are the experts on how their jobs are done. The experiences they have had, the ideas they present and the information they share all help improve the quality of decisions, especially functional decisions such as distance from co-workers and equipment, special lighting or acoustic requirements, and needed furniture elements. Techniques of accessing feedback from occupants help to access this information and ideally need to be part of a phased and planned strategic approach: gathering knowledge about how people work is not the same as asking people to plan their own workspace. Indeed, with too much say in the process, workers tend to favour workspace design that reproduces much of what they already have. Thus the approach to involving occupants must be as carefully designed and managed as the design process itself (Steele, 1986; Becker and Steele, 1994).

There are many ways of environmentally empowering users and not all of them work well in all situations. In many organisations, participation in a space planning process takes too much time away from the job. Not only do managers not welcome these kinds of interruptions, but workers themselves are often not comfortable spending their time on tasks that are not in their job descriptions. Another barrier to empowerment is aversion to conflict. People have different ideas about space and how it should work, and not everyone likes to stand up and express unpopular ideas to their peers. A third barrier is the disempowerment most office workers feel towards workspace. Accustomed to being told where to sit and what kind of space they can have, employees in many companies lack adequate information to make responsible decisions. Workers should be informed so that they can make decisions themselves — not just provided with information about the decisions the design team or CEO is making on their behalf.

9.4.7 Change is positive

Workspace transformation is by nature a positive force. However, it is negative when workspace is changed without attending to its effects on the social order, behavioural norms and work processes, in short, without integrating physical space into a holistic vision of the organisation. The presence of a facilitator can help move such a process forward constructively. When stakeholders are unwilling or unable to resolve or even tolerate conflict, a facilitator should initiate an explicit strategic workspace planning process designed to have buy-in at all levels. Knowing they will have a chance to voice their opinion helps stakeholders be more tolerant of different points of view. In addition, such a process allows decision makers to determine limits on employee empowerment, such as degree of participation, time and cost constraints, and fundamental values and principles to be respected. This ensures accountability and ensures that the process will move forward and will not degenerate.

In view of their experience of building management and operations, and their big-picture understanding of the organisation, individuals with FM training are well placed to facilitate the planning process and to ensure that the full range of user groups and interests are represented. While some training in process facilitation may be necessary (and not all FM professionals will want to take this on), a well-facilitated process can make the difference between a successful and unsuccessful outcome.

Facilitating the user empowerment process helps make workspace decisions more rational while at the same time respecting occupants' emotional and territorial concerns. Conventional design decision making often means that workspace decisions are made on the basis of one person or committee giving approval to plans and specifications presented by design consultants. In such a context, change founders on the shoals of territorial defence, lack of user empowerment and the pressure to revert to old familiar ways of doing things. As a result, workers may protest and even mutiny — sometimes after move-in. This may account for those cases in which extensive time and effort was spent on innovative workspace, and the company floundered or failed shortly after moving into new premises.

The evidence suggests that workspace change is often not managed as an empowering process for employees, and that companies fail to take advantage of the New Workspace Opportunity. In successful examples, workers who have participated, learned, understood and felt empowered by the process have moved into new workspace with an already established sense of ownership and are effective in immediately using the space as a tool and making it work to their advantage. Senior managers see the value of this in better performance, higher morale, more engaged employees and increased competitive advantage.

9.5   RESULTS OF EMPOWERING BUILDING USERS

In view of the major investment that new buildings and workspace is becoming for modern organisations, and of the increasing contrast between the speed of change in the business world and the lengthy terms of commercial leases and a building's useful life, companies today are becoming more aware of the New Workspace Opportunity. Directing and managing workspace change, and taking advantage of the opportunities offered by designing new accommodation, is increasingly falling to facilities managers, who are increasingly in a position to contribute to organisational effectiveness.

The larger and more complex the organisation planning new accommodation, the more urgent decisions makers feel about involving and empowering employees so that they (and their knowledge) contribute to a successful outcome. Companies in recent years who have invested in user feedback and used their employees' specialised knowledge to help make workspace decisions include Google, Muzak, Shell Oil, Hewlett Packard, Bloomberg, New York Times, Pfizer, a wide range of government agencies, and innumerable smaller companies. While different organisations' approaches to empowering users in the context of a user involvement strategy vary, the principle — that of accessing people's specialised knowledge about their work to help make decisions about workspace change — is the same. In fact, in future companies are likely to draw increasingly on their 'human capital' to help make good workspace decisions, with some innovative and inventive results. To do this effectively, FM is teaming up with HR to take joint responsibility for environmental quality at work (Vischer, 2010).

In taking on this role, and acquiring the skills necessary to exercise it, facilities managers increasingly control the degree to which new workspace is linked with improved productivity — not in the old sense of employees being able to produce more widgets faster, but in the more modern sense of a company that reduces staff turnover, is an effective recruiter of new talent, sustains good customer relations, has a clearly recognisable brand, and is quick to recognise and adopt new tools and processes to ensure competitive advantage.

Informing and empowering users to take responsibility for the environment they work in also changes the FM/user relationship from one of providing a service to clients to one of shared responsibility and partnership. Just as organisations increasingly need to invest in their human capital and to provide an environment that encourages creativity, innovation and initiative as well as rapid and accurate task performance, so they need facilities managers to direct and manage their investment in supportive workspace. In the future, FM will need the skills and insight to devise an appropriate change management process, and to manage occupants' involvement in it and empowerment through it.

REFERENCES

Becker, F. and Steele, F. (1994). *Workplace By Design: Mapping the High Performance Workscape.* San Francisco, CA, Jossey-Bass.

Blyth, A. and Worthington, J. (2001). *Managing the Brief For Better Design.* London, UK, Spon Press.

Dewulf, G. and Van Meel, J. (2003). 'Democracy in Design' in R. Best, C. Langston and G. de Valence (eds.), *Workplace Strategies and Facilities Management*, Oxford, UK, Butterworth-Heinemann.

Duffy, F., Jaunzens, D., Laing, A. and Willis, S. (1998). *New Environments For Working*, London, UK, BRE Publications.

Gifford, R. (2006). *Environmental Psychology: Principles and Practice*, 3rd edn., Colville, WA, Optimal Books.

Hamilton, D.K. and Watkins, D.H. (2009). *Evidence-based Design for Multiple Building Types*, New York, NY, J. Wiley & Sons.

Judge, T.A., Thoreson, C.J., Pucik, V. and Welbourne, T. (1999). Managerial coping with organizational change: A dispositional perspective. *Journal of Applied Psychology*, 84(1), 107–122.

Kübler-Ross, E. (1973). *On Death and Dying*, London, UK, Routledge.

Mintzberg, H. (1994). *The Rise and Fall of Strategic Planning: Reconceiving Roles for Planning, Plans, Planners*, New York, NY, Free Press and Prentice-Hall International.

O'Mara, M (2000). *Strategy and Place: Managing Corporate Real Estate and Facilities for Competitive Advantage*, New York, NY, The Free Press.

Preiser, W. and Vischer, J.C. (2004). *Assessing Building Performance*, Oxford, UK, Elsevier Science Publishing.

Sanoff, H. (1999). *Community Participation Methods in Design and Planning*, New York, NY, J. Wiley & Sons.

Steele, F. (1986). *Making and Managing High Quality Workplaces: An Organizational Ecology*, New York, NY, Teachers College Press.

Stevenson, F. and Leaman, A. (2010). Evaluating housing performance in relation to human behaviour: New challenges'. *Building Research and Information*, 38(5), 437–441.

Vischer, J.C. (1996). *Workspace Strategies: Environment As A Tool For Work*, New York, NY, Chapman and Hall.

Vischer, J.C. (2005). *Space Meets Status: Designing Workplace Performance*, London, UK, Routledge/Taylor and Francis.

Vischer, J.C. (2007). The concept of workplace performance and its value to managers, *California Management Review, Winter*, 62–79.

Vischer, J.C. (2010) Human Capital and the Organisation-Accommodation Relationship. Chapter 19 in A. Burton-Jones and J.C. Spender eds. *Oxford Handbook of Human Capital*. Oxford, UK, Oxford University Press.

10 Post-occupancy Evaluation of Facilities Change

Theo J.M. van der Voordt, Iris de Been
and Maartje Maarleveld

CHAPTER OVERVIEW

This chapter discusses possible aims, tools and deliverables of post-occupancy evaluations (POE) (otherwise known as building-in-use studies), with a focus on interventions in supporting facilities. POE has a long tradition and has been applied in different fields (e.g. offices, educational buildings, healthcare facilities, retail and leisure, as well as residential areas). The objectives of POE are various and can include:

- delivering input to an improvement plan
- building up a generic body of knowledge by exploring and testing scientific theories
- developing practical design guidelines and decision support tools.

Data collection tools are also varied, ranging from observations, interviews and web-based questionnaires to walk-throughs and use of narratives. A number of different data analysis techniques are available as well, including qualitative methods, such as content analysis, and quantitative methods, such as descriptive and inductive statistical analyses. A case study illustrates the application of different data collection tools. This case study seeks answers to the effects of new ways of working on employee satisfaction and perceived labour productivity. This study has been conducted by the Center for People and Buildings, Delft, the Netherlands, which specialises in research into the relations between people, working activities and the working environment. The case study is an example of physical interventions including changing the office layout, new furniture, new information and communication technology (ICT) and document storage systems, as well as the flexible use of workplaces. A POE of three pilots has been conducted to test if the new environment performed well as perceived by the managers and employees. The research data has been used in the first instance in order to test if the organisational goals and objectives have been attained and to support decisions with regard to the next steps in this change process. The research data has been used in the second instance more

Facilities Change Management. Edited by Edward Finch.
© 2012 Blackwell Publishing Ltd. Published 2012 by Blackwell Publishing Ltd.

generically, as input to a database for cross-case analyses, exploring and testing hypotheses and benchmarking objectives.

Keywords: Post-occupancy evaluation; Data collection tools; Benchmarking; Decision support; Buildings-in-use.

10.1   INTRODUCTION

Post-occupancy evaluation (POE) is a tool that is being used to investigate users' experiences (satisfaction, perceptions and preferences) and user behaviour in connection to the built environment. In the wider context, including technical and economic issues, a common term is building performance evaluation (BPE) (Preiser and Vischer, 2005). POE goes back to the 1960s and 1970s when there was increasing attention being paid to user participation and user oriented design and management. New disciplines such as architectural psychology and environmental psychology came to the fore (Proshanski *et al.*, 1970; Küller, 1973; Zimring and Reitzenstein, 1980; Bell *et al.*, 2001; Gifford, 1987/2002). Professionals and scientists working in this field started to meet at annual or biannual conferences of the International Association of People-Environment Studies (IAPS) and the Environmental Design Research Association (EDRA). Preiser *et al.* (1988) published their renowned book *Post-Occupancy Evaluation*. Whereas early POEs would focus on buildings and places (and, on a larger scale, residential areas and greenery), later work has also connected with facilities management (FM) as well (e.g. Preiser, 1993; Eley, 2001; Alexander, 2004). Nowadays POE is being applied to many different environments and facilities: for instance, to investigate the added value of FM, workplace management, performance management and sustainability.

Post-occupancy evaluations can be conducted for different reasons and for different target audiences: differing in breadth and depth, method of evaluation, time of evaluation and the people involved in the evaluation. All these points need to be considered when preparing an evaluation. In other words, there must be a clear picture of what is to be evaluated, why, how, when, for whom and by whom (Van der Voordt and Van Wegen, 2005).

10.2   AIMS AND OBJECTIVES OF POE

Evaluation allows lessons to be learnt which can lead to an improvement in the project under investigation and more generally to improve the quality of programming, designing, implementing and managing of facilities. Professionally, the main objective of a POE is to check whether the purposes of a project were achieved or not, to figure out if problems were solved, to make sure that the project briefing was followed in the design and construction phase or to show the client the improvements done comparing the pre-design evaluation with the POE results. Other reasons for the exercise can be both ideological and economic, for example the promotion of health and welfare or a reduction in the facility costs. There can also be scientific goals, such as contributing to the formation of new theories or developing new tools (Table 10.1).

10.2.1   Testing aims and expectations

Stakeholders involved in facility management have all kinds of wishes and expectations with regard to different facilities. The user wants facilities that support their activities effectively

Table 10.1   Goals and objectives of POE.

- To test if client's goals and objectives have been reached
- To record unanticipated results, positive or negative
- To legitimise a continuation or adaptation of accommodation policies
- To steer improvement and upgrading of buildings
- To monitor trends and developments
- To explore and test theories that improve our understanding of complex relationships between facilities, ways of working, organisational needs and user preferences
- To explore and test theories on complex decision-making processes
- To build up a data base, including best practices and worst cases, for theory development and benchmarking purposes
- To deliver input to the strategic brief, project brief, concept and design of new projects (pre-occupancy evaluation)
- To deliver tools, design guidelines and policy recommendations

Source: Slightly adapted from Mallory-Hill, van der Voordt & van Dortmont (2005).

and efficiently, with an attractive 'look and feel'. The client may possibly want the facilities to add value to the organisation in terms of improved productivity, profitability and competitive advantage. This may be achieved by increased production, improved client and customer satisfaction, a positive corporate identity, distinctiveness and a reduction of investment and operating costs (de Vries *et al.*, 2008; Jensen *et al.*, 2010). *Ex post* evaluation establishes whether expectations were fulfilled and whether aims were actually achieved. Besides checking against explicitly formulated aims and expectations, evaluation can also bring to light unintended and unforeseen phenomena, positive and negative. A critical evaluation can give an insight into strengths and weaknesses, opportunities and threats (SWOT analysis) (Hill *et al.*, 1997; Ferrell *et al.*, 1998).

10.2.2   Exploration and testing of theory

Apart from allowing optimisation of the building under evaluation, there are other higher-level arguments in favour of evaluation, above and beyond the individual project. Evaluation makes it possible for others to learn from experiences during the design and construction process and in the use and management phase of similar projects. Individual evaluations and comparisons with other buildings and planning processes can make a significant contribution to the development and testing of theories (for example, on the relationship between facilities or facilities management and human behaviour and experience and the effect on organisational goals and values, or between design decisions and design quality, cost and environmental impact).

10.2.3   Improving understanding of decision-making processes

Decisions are often based on different considerations. The role played by emotions, intuition, judgements and prejudices, social ideals and norms and values is at least as important as that played by rational argument and available information. As such one might speak of 'bounded rationality' (Simon, 1978; Rubinstein, 1998). The evaluation of decision-making processes in facility management can lead to a better understanding of the motives underlying the decisions, roles of the various participants, and use of data and information. Such an understanding is also important for interpreting the result of a product evaluation and guidelines and policy recommendations derived from it. Points requiring attention include the significance of research in decision making, the use of tools, the influence of limiting preconditions and the resolution of conflicting interests. There is also a psycho-logical reason for evaluating facilities or facilities change management processes. Change is exciting, but can involve a good deal of stress. Everyone involved will have spent a good deal of time and energy searching for optimal solutions consistent with the budget, reaching compromises, moving and rearranging. Scheduling an opportunity for evaluation will allow people to express their frustrations, enthusiasms and dissatisfaction.

10.2.4 Database of reference projects

Systematic documentation of the findings of evaluation can lead to the creation of a database of interesting projects, containing a number of key items of information about the project and the findings of the evaluation. Modern relational databases allow the linking of CAD drawing and analytical tools in a networked environment. Box 1 shows the results of an analysis of the CfPB database on relations between satisfaction with facilities and individuals' perceived labour productivity.

10.2.5 Input to existing or new decision-making processes

Findings from *ex ante* POE or pilot studies, as part of a change process, can allow bottlenecks to be identified in good time. Careful evaluation will increase the likelihood of successful decisions and a positive return on investment. Changes are often easier and less expensive in the preparation phase than improvements after implementation. The results of a project-oriented *ex post* evaluation can be used to solve teething troubles and to indicate minor adjustments or radical improvements. Depending on the problems identified, possible solutions might be functional, technical, social or economic (in terms of varying the price/performance ratio). If there is a major mismatch between supply and demand, replacement by other facilities might be the best solution. Lessons learnt from *ex post* evaluations can also be used as an input in *ex ante* evaluation of new projects in order to avoid mistakes and to support evidence-based decision making.

10.2.6 Tools, design guidelines and policy recommendations

Knowledge and understanding are essential preconditions for well-considered decisions, but the results of POE need to be 'translated' into a form which will be quickly and easily accessible to clients, designers, consultants, policy makers, real estate and facility managers and other stakeholders in the building process. Results may be presented in forms such as checklists, design guidelines, seals of approval and manuals. Tools of this kind can be highly effective for formulating and checking building plans, avoiding mistakes, directing policy and developing legislation and regulations.

10.3 DATA-COLLECTION METHODS

In recent decades, a sizeable number of data-collection strategies and methods have been developed, including surveys, case studies and experiments, questionnaires, checklists and assessment scales, individual and group interviews, workshops, walkthrough observations,

Box 1

Findings of the Analysis of the CfPB Database on Relations Between Satisfaction with Facilities and Perceived Labour Productivity (Batenburg and van der Voordt, 2008)

A statistical analysis of data collected with an extended version of WODI Light – one of the tools that has been developed by the Delft Center for People and Buildings (see Section 10.3.1) with 2197 respondents from 17 different office environments, showed a significant but weak correlation between user satisfaction on facilities and self estimated percentage of time that one is being productive. Much stronger correlations came up between satisfaction about facilities and users' perceptions of the supporting impact of the working environment on ones own productivity. In a questionnaire used for this study the respondents were asked to indicate their degree of satisfaction with 63 aspects of the physical working environment. All satisfaction items were measured with a 5-point Likert scale, with $1 =$ highly unsatisfied until $5 =$ completely satisfied. The aspects were categorised in nine sub-dimensions. The most satisfactory sub-dimension was the worksite (4.4), indicating that desks are generally comfortable and ergonomic. The average satisfaction with the climate conditions was relatively low (3.0), a result that resembles earlier research findings.

The perceived productivity of employees was measured in two different ways. Firstly, respondents were asked: 'During what percentage of your working time are you productive?' (Model A). The average response was 78% of total working time, though some respondents stated a much higher percentage (up to 100%) and some a much lower value. Secondly, respondents were asked the extent to which the working environment supported 10 different aspects of their own productivity, such as efficient communication with colleagues and absence of health complaints. The average score here was 3.3 on a 5-point scale, indicating that the respondents were reasonably satisfied with the perceived productivity support. The response to this question was combined with the response to the request of assigning a mark to the degree to which the overall working environment supported ones own productivity (Model B). Here we used the scale which people were accustomed to from school and university (where 6 is a pass, 8 very good and 10 outstanding). The average mark assigned was 6.4, which agrees well with the mean score of 3.3 on a 5-point scale.

The added value of the physical working environment for productivity has been tested with taking into account three other factors that can be expected to influence the (perceived) productivity of office employees: level of job satisfaction, level of satisfaction with the organisation, and personal and job characteristics.

Results

The net relationship between the employee's estimate of his or her own productivity (proportion of total working time spent productively) and facility satisfaction level was showed to be significant, but the regression coefficient was smaller compared to the effect of job satisfaction. Quite remarkably, personal and job characteristics and organisational satisfaction were not significantly related to this measurement of perceived labour productivity. The explanatory power of Model A was relatively low, as the explicative variables only account for 11% of the observed variance in the dependent variable. Model B clearly showed that employees who are satisfied with the facilities rate the degree of support for their productivity provided by the working environment significantly higher. The effect of this factor on the perceived productivity was considerably larger than the effect of job satisfaction, satisfaction with the organisation and personal and job-related characteristics. 54% of the variance in perceived productivity was explained by the four key factors and underlying characteristics considered in our model.

The scatter diagrams (Figures 10.1a and b) relating satisfaction with facilities and perceived (support of) individual productivity (z scores) shows that both models are linear. However, Model A shows a great deal of spread around the theoretical line with its slight positive slope derived from the regression analysis.

Based on the multivariate regression analyses, it can be concluded that the working environment has a fairly limited effect on perceived productivity, especially in relation to the many other factors that were not considered in our model. However, when asking people how satisfied they are about the support of the working environment to being able to perform a number of activities, in particular satisfaction with the facilities, showed to have a substantial influence on perceived productivity. Further analyses in depth revealed that both functional aspects and psychological aspects of the working environment – such as agreeable working surroundings, adequate privacy and inspiring office design –affect perceived labour productivity. It should be emphasised that the focus of the research discussed here was on the relationship between satisfaction and perceived labour productivity, and not on the connection between objective facility performance indicators and actual labour productivity. A review of literature (Van der Voordt, 2003) traced a number of studies showing strong effects of ergonomic furniture, high-quality lighting, noise reduction, design interventions to facilitate team work and the introduction of tele-working on drops in absenteeism, reduction in meeting time, reduction in duplicate files, decrease in errors and higher self-reported productivity.

Perceived own productivity level

Scatter diagram relating satisfaction with facilities and perceived individual productivity

Satisfaction with facilities

Perceived support of productivity

Scatter diagram relating satisfaction with facilities and perceived degree of support for productivity from working environment

Satisfaction with facilities

Figure 10.1 Scatter diagrams of satisfaction with facilities and perceived (support of) individual labour productivity.

and analysis of documents (Lang *et al.*, 1974; Zeisel, 1981/1991; Bechtel *et al.*, 1987; Baird *et al.*, 1996; Vos and Dewulf, 1999; Boardass and Leaman, 2001; Groat and Wang, 2002; Preiser and Vischer, 2004; Van der Voordt and Van Wegen, 2005). Since its inception in 2001, the Center for People and Buildings (CfPB), Delft, the Netherlands, has developed a number of new tools that focus on decision support and pre- and post-occupancy evaluations of working environments (see Chapter 7 for pre-design evaluation). So far the CfPB tools include the WODI toolkit with a set of working environment diagnostic tools (Maarleveld *et al.*, 2009), a workplace guide (Van Meel *et al.*, 2007), a workplace game (De Bruyne and de Jong, 2008), an accommodation choice model (Ikiz-Koppejan *et al.*, 2009), and two tools to deliver quantitative data about places and costs — the PACT tool (Places and ACTivities) to estimate the number of workplaces needed, overall and per type of workplace, and the PARAP lifecycle cost model. Most tools can be used in post-occupancy evaluation and in pre-design evaluations (PDE) as well. We will briefly summarise the tools that have been used in a POE of three pilots in a facilities change process that will be discussed in the next section.

10.3.1   Wodi Light

In order to be able to measure employee satisfaction with the working environment, the CfPB developed the so-called Work Environment Diagnostic Instrument (WODI) (Volker and van der Voordt, 2005; Maarleveld *et al.*, 2009). Later on a shortened web-based questionnaire (WODI Light) has been developed that can be filled out in 10 minutes. The WODI Light questionnaire focuses on issues that turned out to be of utmost importance to overall employee satisfaction and labour productivity. The questionnaire includes a number of thematically clustered questions on a 5-point scale together with several questions on personal characteristics and overall appraisals using a 10-point scale (for an example, see Figure 10.2). Themes include (satisfaction with) organisation, work, the building as a whole, the working environment and workplace, privacy, concentration, communication, document storage, IT, indoor climate, external services and perceived support of labour productivity. The respondents are asked to report their actual use of workplaces and percentage of time spent on different activities during a regular day. The results of the survey can be compared with the average percentage of satisfied and dissatisfied employees in all other WODI Light case studies on a number of key performance indicators (see Section 10.3.2). All data are stored in an ever growing database to be used for further research.

10.3.2   Wodi Light performance indicators: satisfaction and dissatisfaction

Based on the results of a cross case analysis of WODI Light data from over 6500 respondents in 41 cases (19 organisations) conducted in the period 2007–2009, average percentages of satisfied and dissatisfied employees have been calculated, resulting in a list of key performance indicators (Table 10.2). Organisations can use this data to define their own targets for the level of employee satisfaction on issues with a high impact on employees' overall satisfaction, or to compare one's own working environment performance with the perceived performance of other organisations. The WODI Light indicator is based on the average scores on dissatisfaction (marks 1 and 2 on a 5-point scale) and satisfaction

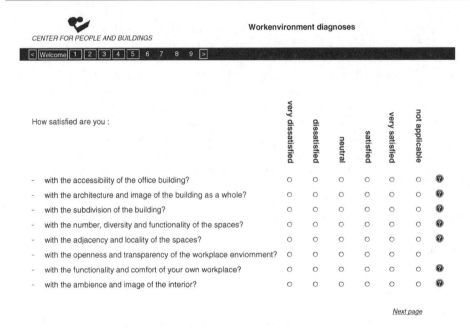

Figure 10.2 Part of the digital WODI Light questionnaire.

(4 and 5); neutral appraisals (3) are not included. The indicator will be updated each year based on additional data that has been collected in the former year. Data of organisations that evaluate their work environment by WODI will be automatically compared with the WODI Light key performance indicators. Instead of setting the goal 'to perform equally or better than the average satisfaction scores' it is also possible to use the indicators and the ranges as a reference to decide on another aimed percentage of satisfied employees per aspect, in order to match with organisational objectives and constraints. Another possibility is to use the underlying WODI data to calculate a new benchmark indicator. For example, an organisation could decide to strive for percentages of satisfied employees that are equal or higher than in the three best-performing buildings. A third possibility is to choose an 'a priori' standard, e.g. 'at least 80% of the employees should be satisfied with the working environment'.

According to Table 10.2, a high percentage of Dutch employees are satisfied with the content and complexity of their work, the accessibility of the building, and support of communication and social interaction. On the other hand, many employees expressed dissatisfaction with the possibilities for concentration, privacy and indoor climate. It should be emphasised here that the range per item was quite large.

10.3.3 Workplace Game

The Workplace Game is a communication tool that enables office workers to exchange ideas about the use of the office environment through open discussion. It makes often implicit thoughts about behaviour in the office more explicit. Playing the game entails walking through an imaginary work environment with colleagues while facing and discussing several situations with regard to values and norms, information and knowledge, and attitude and behaviour in the work environment. Depending on the position of the organisation — for

Table 10.2 Key performance indicators 2010 — showing the average percentage of satisfied and dissatisfied employees based on 41 WODI Light cases and the range between minimum and maximum percentages (Brunia et al., 2010). Reproduced by permission of the Centre for People and Buildings.

Aspects of the workplace environment	Satisfaction		Dissatisfaction	
	Indicator 2010	Range (min — max)	Indicator 2010	Range (min — max)
Organisation	64%	41–86%	11%	0–30%
Content and complexity of work	79%	40–100%	6%	0–29%
Sharing own ideas regarding work environment	43%	7–66%	22%	0–64%
Accessibility of the building	77%	51–96%	12%	1–32%
Architecture and 'look' of the building	53%	8–96%	20%	0–69%
Subdivision of the whole building	47%	18–80%	23%	5–51%
Number, diversity and functionality of the spaces	45%	15–77%	25%	0–52%
Adjacency and locality of the spaces	54%	27–80%	18%	6–42%
Openess and transparency	54%	30–86%	19%	3–41%
Functionality and comfort of own workspace	59%	30–82%	21%	0–42%
Ambience and 'look' of the interior	54%	15–88%	21%	0–57%
Privacy	37%	10–79%	37%	9–75%
Possibilities for concentration	40%	15–86%	38%	14–73%
Communication and social interaction	70%	44–92%	11%	0–34%
Archive and storage facilities	35%	11–71%	29%	12–60%
ICT and supporting services	55%	29–95%	18%	0–39%
Facilities and facilities management	55%	31–70%	11%	2–28%
Indoor climate	40%	22–61%	35%	16–48%
Lighting	62%	41–85%	14%	2–28%
Acoustics	46%	22–69%	26%	8–50%
Possibilities for remote working	42%	5–89%	20%	2–55%

example, prior to a renovation or renewal of the present building or a move to another building or when managing a new environment — the Workplace Game can be used to create discussions about new rules and regulations for behaviour. It can also be used to stimulate shared values and norms, to create awareness of the (impact of) workplace change, to stimulate the preferred use of the work environment and to raise awareness of one's own points of view, as well as suppositions and norms in relation to the work environment.

10.3.4   Space utilisation monitor (SUM)

To measure the actual use and occupancy of the work place, the CfPB developed a software application for a handheld computer. During the walk through the researcher records whether the workplaces are vacant, temporarily vacant or occupied and if 'yes' also the performed activities. Usually this involves eight measurements a day during a working week. The two days that show the highest average occupancy will be measured again in the following week. The provided output includes charts of occupancy levels and activities per type of workplace, per hour, per day or per department. These data can be used to support decision making about the introduction of hot-desking, the ratio of the number of desks to the number of employees, and the number of workplaces per type of workplace (open setting, places for concentration, informal and formal meeting places and so on) (Maarleveld *et al.*, 2009).

10.4   APPLICATION IN PRACTICE: A CASE STUDY

10.4.1   Context and aims of the case study

To illustrate the use of several data collection tools, a case study will be discussed that focuses on employee satisfaction with regard to the work environment. The main tasks of the public educational organisation concerned are academic examination, administration and customer service. This organisation is facing several changes in both the organisational structure and the working environment. As a consequence of merging with another organisation the strategy, vision and working methods will change as well. Due to ICT developments, face to face contacts will increasingly be replaced by virtual contact via the Internet. The organisation will move into a new office building in 2011. The challenge is to develop and create a new work environment that supports the (new) work methods and processes. The organisation aims to achieve a better, more pleasant and less expensive housing solution and has opted for a new building and a well-considered design and implementation process in order to elaborate these goals. The process has been set up with a high level of employee and management participation. The case study included reflections on the new office concept and the implementation process (based on insights from earlier research), an *ex ante* evaluation of the present environment and *ex post* evaluations of three pilots in which the future workplace concept has been applied during one year to test if it works well (De Been and Maarleveld, 2008). With this case study the organisation aimed to use the pilots as a 'living lab':

- to get insight into future processes, preferred behaviour and working methods of employees and management
- to help employees and managers to gain insight into the consequences of new working methods and working processes for the office concept
- to learn from these preliminary experiences and to use the lessons learned as input to the next phase, i.e. the design and implementation of the new office concept in the whole organisation.

The pilot environments also functioned as an example for all users of the future office building, providing them with an opportunity to familiarise themselves with the new office concept. The members of the overall project group and the three pilot working groups were strongly involved in the research process. The overall project group was responsible for the development and implementation of the new office concept for the whole organisation. The pilot working groups represented the pilot departments and were responsible for the provision of information and the involvement of their colleagues. Pilot group A did so weekly; the working group kept their colleagues informed and asked for their opinion and input when they thought this was necessary. The other two working groups informed their colleagues on a monthly basis, but did not ask for their opinion. One of groups B and C could only start informing their colleagues halfway through the design phase as it was unknown to them which group was going to move into the pilot area.

10.4.2   Data collection

The first step of the research was to get insight in the strategy, aims and constraints of the organisation and of the pilots in particular. Information concerning roles, research approach, aims and restrictions with regard to the office concept and communication was been collected by interviews with key players of the organisation and a workshop with the working groups (Figure 10.3). This not only brought up valuable and sound research information but it also made the organisation more aware of their aims and constraints, prompting them to formulate their aims more clearly and explicitly.

Two months later the project group and pilot working groups made a field trip to two different office buildings, both with an innovative office concept, to make them aware of the opportunities and risks of the new office concept. Hereafter a 'zero measurement' has been conducted using the WODI Light tool in order to get insight into employee satisfaction, perceived influence on productivity and the actual use of the 'old' working environment present in the pilot departments.

Two out of three pilot departments had formulated 'employee satisfaction' as one of the objectives. In several workshops the WODI Light results were discussed regarding employee satisfaction and perceived productivity, whilst at the same time developing

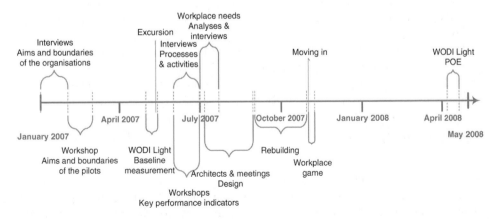

Figure 10.3   Research activities in chronological order.

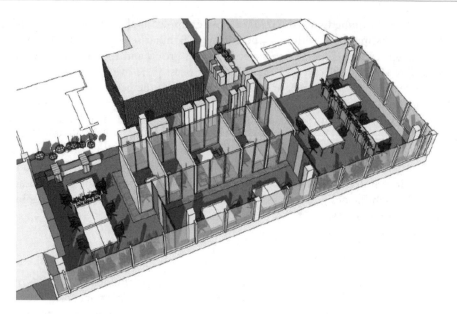

Figure 10.4   Final design of one of the new working environments (pilot A).

their own sought-after performance level. Various sources of information have been used for this, such as (1) the WODI Light key performance indicator (see Section 10.3.2); (2) data from an earlier case study where an adjusted performance indicator had been developed; (3) the aims of ones own pilot department; and (4) the results of the WODI Light zero measurement.

To make an evidence-based proposal of required numbers and types of workplaces, the distribution of activities that took place during work time was thoroughly discussed in the pilot working groups and, where necessary, adjusted. Subsequently, five interior architects presented possible solutions for the office lay-out and interior design. Eventually, three out of these five architects were assigned to one of the pilot departments to work on the final pilot design. The architects were handed all information that came out of the research with regard to the distribution of activities and satisfaction with the present working environment. Figure 10.4 shows the final design of one of the pilot departments (A).

The new pilot designs differ from the old design by application of a huge diversity of activity-based workplaces, a reduced total work area (by flexible and shared use of workplaces instead of personal desks), increased openness and transparency, new IT facilities, a shift to digital archiving and less individual space to store documents or other material.

10.4.3   Moving in

Just before and during moving into the new office environment the employees of the three pilot departments discussed several issues regarding the new environment by playing the Workplace Game (Maarleveld, 2008). It appeared that the attitude towards the new design differed a lot between the groups. Employees of the two pilots who were only informed monthly were somewhat sceptical towards the new design and new ways of working whereas others were much more positive. The discussions during the game stimulated all three pilots to formulate regulations concerning the use of the new work environment, including rules about not eating at a workplace and applying a clean-desk policy when leaving it for more than two hours. Not all new working environments were finished off completely when the pilot groups moved in. It took one to two weeks before everything was fully arranged.

10.4.4   Post-occupancy evaluation

Four months after moving into the new pilot environments, a post-occupancy evaluation was carried out using the WODI Light tool and the Space Utilisation Monitor (SUM). Using the same WODI Light tool before and after the move made it possible to compare the data from the zero measurement before the move and the post-occupancy evaluation of the new environment. The SUM tool provided detailed information about occupation levels and use of different types of workplaces in the new work environment and gave insight in the suitability of the new office layout and the number and types of workplaces.

Figure 10.5 shows a comparison of the zero measurement *ex ante* and the post-occupancy evaluation of pilot A with regard to employee satisfaction in relation to several aspects of the working environment. In the new situation many aspects of the physical environment were being appraised more positively. These included the ambience and look of the interior, openness and transparency of the work environment, functionality and comfort of the workspaces. However, some other aspects are now being judged more negatively (e.g. facility management, archive, storage facilities and ICT).

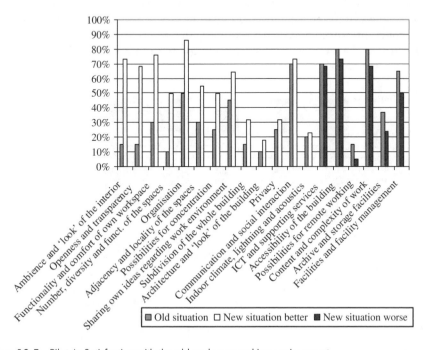

Figure 10.5   Pilot A: Satisfaction with the old and new working environment.

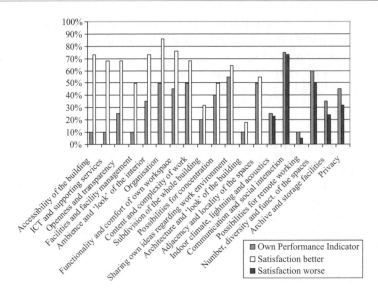

Figure 10.6   Pilot A: satisfaction with the new environment compared to predefined satisfaction levels.

The pilot A department also compared the post-occupancy evaluation data with the adapted WODI key performance indicator based on their discussions about aimed satisfaction levels, the results of the WODI Light zero measurement and satisfaction levels in a particular case study (see Figure 10.6). With the new environment the pilot A group seems to have achieved their pre-set goals adequately, except for privacy, archive and storage facilities as well as the number and diversity of workspaces.

The findings from the workspace occupation measurement showed that the variety and number of workspaces generally meet the needs of the pilot A department (Figure 10.7). The occupancy level was almost the same for the different types of

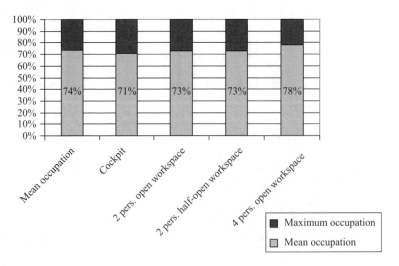

Figure 10.7   Average and maximum occupancy levels of workspaces pilot A.

spaces, which may indicate that most spaces are equally popular. However, though the mean occupancy level is 74%, the maximum occupancy level of all spaces together was 100%. So, on at least one occasion all workspaces were occupied, leaving no space for those who sought a workplace.

10.4.5 Lessons learned

Different data collection tools (interviews, workshops, workplace game, web-based questionnaire) were used to measure employee satisfaction in the old situation and after the move (post-occupancy evaluation). Space utilisation measures were also considered in the before and after situations. The results provided extensive information and points of special interest to assist in the preparation and implementation of a new office concept. The pilot study showed that informing and involving the users in the process is of utmost importance. Key players not only should inform their colleagues about the goals and objectives of change and the planning of the implementation process, but should also ask for and deal with their opinions. Field trips to other innovative offices and visual presentations of the future office concept proved to be very useful. Furthermore, it turned out that the management played an important role in reducing resistance and creating employee support. They served as an inspiring example to others by, for example, apply a clean-desk policy and sharing workplaces themselves.

The post-occupancy evaluation showed that the overall satisfaction with the new office environment was higher compared to the former office environment. This was particularly so with regard to the architecture and interior design of the work environment, functionality and comfort of the workplaces and openness and transparency of the work environment. The space utilisation measurement showed that a high proportion of different workspaces were used. However, other aspects were evaluated more negatively following the move into the new office environment. This applied to supporting facilities such as archive and storage facilities and facility management. This finding emphasises the importance of providing sufficient and well-designed supporting facilities, including ICT, to support a new flexible working environment. Providing guidance and support after the move proved to be important as well, for instance, by establishing a working group that was responsible for assisting employees and managers in using the new office concept.

Long-term monitoring and continuous evaluation of the new working environment may help to detect inadequacies and suggest ways to address shortcomings, for example in providing a sound understanding of suitable interventions in the new workplace.

10.5 CONCLUDING REMARKS

This chapter discussed possible goals and objectives of post-occupancy evaluations (POE) as well as tools to collect and analyse data that can be used to support decision making, before and after a move. Knowledge from POE research can be used as input into change management processes, both with regard to lessons learned about performance and does and don'ts in managing the process. This was illustrated by a case study of the initiation, design, implementation and management of a new working environment. Application of POE in a number of different cases revealed a huge amount of data that can be used for purposes of benchmarking and building a growing body of knowledge about critical success factors.

Certain 'hard' factors showed to be key to performance, such as:

- sound insulation between places for communication and concentration
- a well considered capacity of archive and storage space
- advanced ICT facilities that work without any problems
- an attractive indoor climate.

But 'soft' factors were also found to have an important role, such as:

- a pleasant architectural appearance of the building and its interior
- psychological issues such as privacy and personal control

The case studies confirmed the findings from literature reviews with regard to managerial conditions for arranging a thorough initiation, design and implementation process and good after care (Table 10.3).

Table 10.3 Critical success factors in accommodation change processes (Van der Voordt, 2003). Reproduced by permission of the Centre for People and Buildings.

1 A meticulous analysis of the organisation and its work processes beforehand.
2 Clear objectives.
3 Adequate project organisation with transparent task agreements and clear authorisation.
4 Commitment of management.
5 Adequate involvement and careful coordination between facility management, real estate management, human resource management and ICT specialists.
6 A well structured implementation process, with an enthusiastic initiator, a balance between top-down and bottom-up management, sufficient information and communication, and enough time for discussion and reflection.
7 Taking opposition seriously, particularly the tension between flexible working in open spaces and human needs for privacy, personal territory, identity, personalisation and status.
8 Providing assistance to employees (i.e. training courses in flexible working and central and digital filing systems).
9 Careful management of the building-in-use including 'guarding' the concept.

Many questions remain unanswered in relation to POE as a general methodology. What were the costs and benefits of facilities change, both in monetary and non-monetary terms? What were the main priorities when taking into account employee satisfaction, productivity, organisational performance, competitive advantage and — of utmost societal relevance — sustainability? How can POE findings contribute optimally to the decision-making process, taking into account the different roles and interest of various stakeholders? Furthermore, how can POE findings be conveyed in an accessible, simple and timely manner? How can they also be linked to common phases in decision making such as setting managerial objectives, searching for alternatives, comparing and evaluating alternatives, making choices, implementing decisions and follow-up and control (Harrison, 1996)? In order to be able to answer at least some of these questions, the Center for People and Buildings is working on further elaboration and testing of the so-called 'Accommodation Choice' model (Ikiz-Koppejan *et al.*, 2009). This is a process model containing four steps to support the initiation, design and implementation of new working environments. Another 'work in progress' activity is a study on how to improve employee satisfaction and how to prevent dissatisfaction in connection with labour productivity. Other interesting next steps could be to compare data and tools from different countries in order to identify contextual and cultural influences, and to undertake a closer examination of the impact of services such as the reception desk, catering, cleaning and security on people's well being and organisational performance.

REFERENCES

Alexander, K. (2004). *Facilities Management*. Innovation and Performance. London, Spon.

Baird, G., Gray, J., Isaacs, N., Kernohan, D. and McIndoe, G. (1996). *Building Evaluation Techniques*. New York, McGraw-Hill.

Batenburg, R., and van der Voordt, D.J.M. (2008). *Do Facilities matter? The influence of facilities satisfaction on perceived labour productivity of office employees*. Paper presented at EFMC 2008, European Facility Management Conference, Manchester.

Bechtel, R., Marans, R. and Michelson, E. (1987). *Methods in Environmental and Behavioral Research*. New York, Van Nostrand Reinhold.

Been, I. de and Maarleveld, M. (2008). *Pilots procesgericht huisvesten*. Delft, Center for People and Buildings. [Pilots process-oriented accommodation].

Bell, P.A., Greene, T., Fisher, J. and Baum, A.S. (2001). *Environmental Psychology*. Fort Worth, Harcourt College Publishers, 5th edition.

Boardass, B., and Leaman, A. (eds.) (2001). Post-occupancy evaluation. *Building Research & Information*, 29 (2), special issue.

Brunia, S., de Been, I., Beijer, M. and Maarleveld, M. (2010). Imagineering van de werkomgeving! *FM Executive*, September 2010, pp. 23–26. [Imagineering the working environment].

de Bruyne, E. and de Jong, A. (2008). *The Workplace Game: exploring end users' new behaviour*. Paper presented at the AKFEI08 Conference, Las Vegas.

Eley, J. (2001). How do post-occupancy evaluation and the facilities manager meet? *Building Research & Information*, 29(2), 164–167.

Ferrell, O., Hartline, M., Lucas, G., and Luck, D. (1998). *Marketing Strategy*. Orlando, FL, Dryden Press.

Gifford, R. (1987/2002) *Environmental Psychology: Principles and Practice*. Boston, Optimal Books.

Groat, L. and Wang, D. (2002). *Architectural Research Methods*. New York, Wiley.

Harrison, E.F. (1996). A process perspective on strategic decision making. *Management Decision*, 34(1), 46–53.

Hill, T. and Westbrook, R. (1997). SWOT analysis: It's time for a product recall. *Long Range Planning*. 30(1), 46–52.

Jensen, P.A., van der Voordt, Th., Coenen, C., von Felten, D., Lindholm, A.L., Balslev Nielsen, S., Riratanaphong, C. and Schmid, M. (2010). *The Added Value of FM: Different Research Perspectives*, EuroFM: Conference paper from the European Facility Management Conference, Madrid, 1–2 June 2010.

Ikiz-Koppejan, Y., van der Voordt, Th. and Hartjes-Gosselink, A. (2009). Huisvestingskeuzemodel: procesmodel voor mens- en organisatiegericht huisvesten, Center for People and Buildings, Delft. [Accommodation choice model to support accommodation processes].

Küller, R. (ed.) (1973). *Architectural Psychology*. Stroudsburg, Dowden, Hutchinson and Ross.

Lang, J., Burnette, C., Moleski, W. and Vachon, D. (1974). *Designing for Human Behaviour: Architecture and the Behavioural Sciences*. Stroudsburg, Dowden, Hutchinson & Ross.

Mallory-Hill, S., van der Voordt, Th.J.M. and van Dortmont, A. (2005). *Evaluation of innovative workplace design in the Netherlands*. In: W.F.E. Preiser and J.C. Vischer (eds.), *Assessing Building Performance*. Oxon, UK, Elsevier, pp. 160–169 and 227–228.

Maarleveld, M., Volker, L. and van der Voordt, T.J.M. (2009). Measuring employee satisfaction in new offices: the WODI toolkit. *Journal of Facilities Management*, 7(3), 181–197.

Maarleveld, M. (2008). *Evidence-based Workplace Design and the Role of End-user Participation*. RAMAU Conference, Paris.

van Meel, J., Martens, Y., Hofkamp, G., Jonker, D. and Zeegers, A. (2007). Werkplekwijzer. Ingrediënten voor een effectieve werkomgeving. Delft: Center for People and Buildings. [Workplace Guide – Building stones to an effective work environment].

Preiser, W.F.W.E. (1993). *Professional Practice in Facility Programming*. New York, Van Nostrand Reinhold.

Preiser, W. and Visscher, J. (eds.) (2004). *Assessing Building Performance*. Oxford, UK, Elsevier.

Preiser, W.F.E., Rabinowitz, H.Z., and White, E.T. (1988). *Post-occupancy Evaluation.* New York Van Nostrand Reinhold.

Proshanski, H., Ittelson, W. and Rivlin, L. (1970). *Environmental Psychology: Man and his Physical Setting.* Holt, Rinehart and Winston.

Rubinstein, A (1998). *Modeling Bounded Rationality.* Cambridge, Massachusetts, MIT Press.

Simon, H.A. (1978). *Rational decision-making in business organizations.* Nobel Memorial Lecture, 8 December, Pittsburgh, Pennsylvania, Carnegie-Mellon University.

Volker, L. and van der Voordt, D.J.M. (2005). An Integral Tool for the Diagnostic Evaluation of Non-Territorial Offices. In B. Martens and A.G. Keul (eds.), *Designing Social Innovation. Planning, Building, Evaluating.* Göttingen: Hogrefe & Huber Publishers, pp. 241–250.

Van der Voordt, D.J.M. (2003). *Costs and Benefits of Innovative Workplace Design.* Delft, Center for People and Buildings.

van der Voordt, D.J.M. and van Wegen, H.B.R. (2005). *Architecture in Use. An Introduction to the Programming, Design and Evaluation of Buildings.* Oxford, Elsevier, Architectural Press.

Vos, P.G.J.C. and Dewulf, G.P.R.M. (1999). *Searching for Data. A Method to Evaluate the Effects of Working in an Innovative Office.* Delft, Delft University Press.

de Vries, J.C., de Jonge, H. and van der Voordt, Th.J.M. (2008). Impact of real estate interventions on organisational performance. *Journal of Corporate Real Estate,* 10(3), 208–223.

Zeisel, J. (1981/1991) *Inquiry by Design. Tools for Environment Behaviour Research.* Cambridge University Press.

Zimring, C. and Reitzenstein, J. (1980). *Post-Occupancy Evaluation: An Overview.* Environment and Behaviour, 12(4), 429–450.

Research findings are also presented and discussed periodically at international conferences, among others:
- the annual European Facility Management Conference (EFMC) (see for instance www.efmc2010.com) that is being organised jointly by IFMA (www.ifma.org) and EuroFM (www.eurofm.org)
- the conferences of CIB W070 (www.fmresearch.co.uk), an international community dedicated to the furtherance of facilities management research
- the biannual conferences of International Association People-Environment Studies (IAPS: www.iaps-association.org)
- the annual Environmental Design Research Association conference (EDRA: www.edra.org).

For more information about the work of the Center for People and Buildings, the Netherlands, see www.cfpb.nl.

11   Change and Attachment to Place

Goksenin Inalhan and Edward Finch

CHAPTER OVERVIEW

This chapter considers the emotional effect of place attachment and the implications for change implementation. With the advent of mobile working and the 'work anywhere anytime' culture we are often led to believe that people are immune to the stress that arises from physical change. This chapter identifies the nature of place attachment and the implications for facilities managers seeking to overcome resistance to change. Understanding exactly what we 'treasure' in workplaces can go some way to overcoming change resistance.

Keywords: Place attachment; Change implementation; Mobile working; Resistance to change; Familiarity; Territoriality in the workplace; Place disruptions.

11.1   THE AGE OF EVERYTHING

Change is an inevitable facet of human existence. People have always had to adapt to changing environmental conditions. In the nomadic hunter gatherer society of early man, there was a constant need to move from one place to another, whilst adapting to different seasons and hostile environments (Natarajan, 2003). However, this adaptive capability is seriously tested by the rate of change modern society is now witnessing. Toffler (1970), an early commentator on the impact of technology on human behaviour, claimed that this accelerated rate of technological and social change can overwhelm people, leaving them disconnected whilst suffering from 'shattering stress and dis-orientation'. This phenomenon was encapsulated in the phrase 'future shock'. Toffler suggested that people's sensing mechanisms for coping and adaptation for 'the prema-ture arrival of the future' were being severely challenged, leading to 'future shock'. He suggested that this accelerative thrust has personal and psychological, as well as sociological, consequences.

In his book '*The Corrosion of Character*' Richard Sennett also portrays an unsettling picture of today's economy. He argues that people have to cope with new concepts of flexibility, flexitime, teamwork, delayering and ever-changing working conditions that are seemingly presenting new opportunities of self-fulfilment to workers, but in reality creating new forms of oppression, ultimately disorienting individuals and undermining their emotional and psychological well-being (Natarajan, 2003).

Facilities Change Management. Edited by Edward Finch.
© 2012 Blackwell Publishing Ltd. Published 2012 by Blackwell Publishing Ltd.

We can no longer assume that things will stay as they have been — salaries, benefits, people, locations, company names, job descriptions and job security are all susceptible to swift and drastic change. The new 'psychological contract' between employer and employee does not provide workers with assured continuity in a company. Increasingly, organisations are judging job continuation on the basis of performance and short-term needs, rather than by 'family-like' emotional ties (Jeffreys, 1995). Though Sennett does not propose any clear solution to this situation, he highlights the necessity of people developing communities to build up their identities.

Knoke (1997) asserts that we are entering the Age of Everything—Everywhere, where the effortless flow of people, products and knowledge occurs from one location to another, enabled by developments in communications and transportation technologies. This is creating a fourth dimension that transcends time and three-dimensional spaces. We are living in a 'placeless' society where 'place no longer matters' and 'everything and everybody is at once everywhere'. However, Knoke does not suggest that society exists in isolation or without anchors: rather, that we exist in a super-connected society where distances ceases to exist, where you can reach about and touch everyone in the world.

The debate about the possible benefits and risks of these changes is ongoing. At this early stage in the introduction of new ways of working, the possible advantages and disadvantages are not fully understood nor has the claimed contribution of flexible working to business performance been thoroughly tested. Some hold a positive view of the potential benefits of flexible working, others adopt a more conservative and questioning approach. However, both camps recognise that changes of this kind demand radical facility solutions (Nutt and McLennan, 2000).

Change has become a way of life for organisations as well, as the business environment has become increasingly competitive. A growing number of organisations are introducing 'new ways of working' in their physical work environments to better respond to dynamics in the work society and handle space (facilities) more effectively and efficiently. Changes in the physical workplace have been used as a catalyst by many organisations to introduce many elements of organisational change. However, the uncertainties associated with such changes can have a deleterious effect on performance and can cause long-term damage to work relations, particularly when little thought is given to the ways such changes are viewed by employees (Mazumdar, 1992; Inalhan, 2006).

11.2   LOSS AND GRIEF

Jeffreys (1995) points out that perhaps the most overlooked consequences of the turbulent changes in the corporate world since the 1980s are the human factors of loss and grief. How does change in the workplace result in 'loss' in the workplace? He claims that whatever we left behind after we have gone through a transition represents loss. Even if the 'new' is desired, we still lose the 'old', and the reaction to this loss is grief. Workplace change brings about a real sense of death (Jeffreys, 1995; Milligan, 2003). The actual loss of people, work routines, location or control disrupts an important part of the person's total identity and meaning in life. The sense of knowing what to expect from the organisation and what is expected by the organisation is also lost, as new psychological contracts are forged in a shrinking work environment (Jefferys, 1995). Survivors of layoffs and reorganisations react with feelings akin to those of people who are grieving.

Jeffreys (1995) further considers the situation where a company alters its organisation by reducing or redeploying its workforce, or makes other substantial changes in the way it meets its goals: a sense of loss of what used to be is generated for employees. Employees grieve the loss of people with whom they have bonded, status in the organisation, a sense of control over their work, familiar procedures and workspace, trusted reporting relationships, certainty about their future and their own assumptions about what could be expected from the company (ASTD, 1990).

While organisations are fairly successful in managing the mechanics of change, $3 billion a year in losses is suffered by companies in the US because of the effects of negative attitudes and behaviours of employees towards workplace changes (Topchik, 2001). Although much has been written about the effects of organisational change on organisations and aspects of individuals' behaviour, the role of the workplace in supporting and influencing the change process is a grey area. While 'managing change' studies provide insights into the effective management of organisational change, they contribute very little to an understanding of the relationship between people and workplace design (physical environment) and change management. The dynamics of transition and the meanings that people place on their workplace are often over-looked in employees' resistance to change.

Grimshaw and Garnett (2000) highlight the lack of debate and research on workplaces as facilitators of work: furthermore, there is a paucity of understanding in relation to its contribution to the success of the organisation and the role played by employees' contributions to the shaping of the work environment, whether individually or as group. Although the individual employee has been the subject of much recent research work, very little is understood about how employees function in their work environment after many changes have been applied to their physical environment. Nor is there an understanding of how their previous bonds and habitual ways of doing things in their old workplace influence the way they function in their new workplaces.

A person's relationship to a physical environment is complex and difficult to understand. However, the key aspect of this relationship is emotional. Psychologists and others concerned with work behaviour have long been interested in employees' feelings in terms of outcomes such as satisfaction, stress and fatigue. In contrast to research on the expression of emotion, research on the experience of emotion is relatively underdeveloped (Briner and Totterdell, 2002). Even though recent interest in the workplace has been intense, the opportunities and challenges remain on the emotional significance of the physical environment for employees.

Today the business world is witnessing the diversification of work practices, more responsive working arrangements, the global dispersal of work, and new multi-venue and multi-location ways of working. These developments are part of the fashionable notion of 'flexible working': work that is 'time flexible', 'place flexible' and 'location variable' (Nutt and McLennan, 2000). Particularly with the rise of 'knowledge work' and more creative forms of office work, the traditional design concepts of the office are being questioned. The idea of one individual within the office, or of one desk per person, is now continuously being challenged.

While the office of the future requires designers, facility managers and corporate real estate managers to think in terms of space and time, rather than about desks and chairs, employees still continue to act upon the old person/place metaphor. The symbolism attached to place is a powerful force that works against locational flexibility. It is argued that many organisations' attempts to implement new workplace strategies have failed due to overwhelming employee resistance to change (Stegmeier, 2008).

The flexibility gained from 'address free' working environments often assumes a 'no-pain' impact on users. However, the acceptance of such workplace strategies depends largely on human nature and the psychological factors related to it. Place attachment presents a challenging view of the world that is contrary to all the received wisdom in facilities management.

Often the place attachments held by employees to the physical site of an organisation go unrecognised by the management involved in such transition processes (Milligan, 2003). Workplace grief (where change occurs) is not given much validity in our society or in business organisations. As a result, feelings are further blocked by business organisations that view employee expressions of negative feelings or even confusion about workplace change as exhibiting a bad attitude (Jeffreys, 1995). Ignoring the emotional charge of these employees not only has the potential to undermine the success of the change management project and the organisation itself, but can also damage the well-being of employees through alienation.

In the knowledge that place attachment is a significant part of human well-being and psycho-cultural adaptation to an environment facilities management has a crucial role. Understanding the natural and expected flow of human reactions as employees move from the old to the new will enable managers to intervene with clarity and compassion. There is a transition time between the old and the new, when people feel adrift, lacking much needed anchors and grounding for people those undergoing change. Special attention must be given to the grief process during this time. Ignoring this grieving process will extend the period of human pain and of lowered productivity (Jefferys, 1985).

Therefore this chapter will address the issues of employees' 'perception of change' from the socio-psychological and behavioural point of view, and the consequences of this for the employee's ability to adopt to new environments. Following the literature on place attachment, the next section discusses the findings of work by Inalhan and Finch (2004) in relation to place attachment in the workplace. The 'place attachment in workplaces' framework is introduced. The chapter will end with a reflection on the ramification of these findings on facilities management.

11.3 IS PLACE ATTACHMENT HEALTHY?

Altman and Low (1992) define *place* as an environment that has been given multiple meanings through personal, group or cultural processes, all of which are important for health. The psychology of place is based on the assumption that people strive for a sense of belonging to a place (Fullilove, 1996).

A sense of belonging is necessary for psychological well-being and depends on strong, well-developed relationships with nurturing environments. Psychiatrists claim that high levels of mobility has clear psychological costs for adults and children (Fullilove, 1996).

A sense of belonging arises from the operation of three key psychological processes between people and environment.

1. Attachment: the processes conceptualised as a series of emotions and behaviours that modulate distance from, and hence maintain contact with, the object of attachment, which is a source of protection and satisfaction.

2. Familiarity: the processes by which people develop detailed cognitive kno⌐
   environs. This accumulated knowledge is essential for survival.
3. Identity: the processes concerned with the extraction of a sense of ⌐
   the places which one occupies in life. Sense of spatial identity is fu⌐
   human functioning.

Displacement breaks these emotional connections. The ensuing disorientation, nostalgia and alienation may undermine the sense of belonging and mental health in general (Fullilove, 1996). Freeman (1984) emphasises in his edited book '*Mental Health and the Environment*' that too much change is detrimental to human life and ought to be avoided. Freeman's psychological conservation of environment theory rests upon the benefits of stability which allows people to develop intimate knowledge of their settings and to develop trusting relationships with place and with each other. Fullilove (1996) argues that the most serious threat to human well-being is the disintegration of communities, which can both precede and follow large-scale displacements (including those driven by technology and resizing).

With the emergence of the rapidly changing global economy, situations that undermine our ability to form attachments with people, places and companies demand attention in relation to the human cost of change.

This view is also expressed by Sennett in '*The Corrosion of Character*' as:

Recent developments in the working conditions of modern work life have brought increased alienation from work and family, much less job security and a new set of dominant values which elevate the individual and destroy the ties of co-operation, not to mention solidarity.

However,

... one of the unintended consequences of it is that it has strengthened the value of place, aroused a longing for community ... for some other scene of attachment and depth. (Sennett (1999, p. 138))

Sennett draws conclusions about the personal consequences of work in the new capitalism (the 'new work order' as James Gee called it). He argues that people have to cope with new concepts of flexibility, flexitime, teamwork, delayering and ever-changing working conditions that are seemingly presenting new opportunities of self-fulfilment to workers, but in reality are creating new forms of oppression — ultimately disorienting individuals and undermining their emotional and psychological well-being. Place attachment reflects a community desire to do something about the erosion of the glue that holds society together (Wood and Giles-Corti, 2008).

The workplace is a particularly worthy context in which to examine this process of change and transition in relation to place attachment, since the frequency with which changes such as relocations and renovations occur in these settings continues to increase (Fisher and Cooper, 1990). However, there is limited knowledge and understanding on the subject of place attachment and the psychological impacts of the changed work environments on employees. The place attachments held by employees to the physical site of an organisation go unrecognised by the management involved in such transition processes (Milligan, 2003; Inalhan, 2009).

In the knowledge that place attachment is a significant part of human well-being and psycho-cultural adaptation to an environment, place attachment has a crucial role in facilities management. This chapter is about employees' 'perception of change' from the socio-psychological and behavioural point of view, and the consequences of this for the employee's ability to adopt new environments; the relationship between physical environment and social and environmental psychology has been the central issue. It illuminates ways in which isolated employees, previously prevented from achieving social reintegration, can re-establish the conditions for attachment and a sense of belonging.

11.4 DIMENSIONS OF PLACE ATTACHMENT

Place attachment can be defined as one's emotional or affective ties to a place and is generally thought to be the result of a long-term connection with a place (Low and Altman and Low, 1992). Milligan (1998) describes this emotional tie as being formed by an individual to a site as a result of the meaning given to the site through interactional processes. She suggests that such attachment is comprised of two interwoven components: 'the interactional past' and the 'interactional potential' of the site.

The interactional past refers to past experiences: in other words 'memories' associated with a site. Places have the power to recall emotions and stir memories that have been dormant while the person was away from the place (Milligan, 1995).

Interactional potential refers to the future experiences imagined and anticipated to be possible in a setting, or in other words 'expectations'. An individual's experiences within and in relation to a specific site results in a set of expectations for future interactions in the site (Milligan, 1995).

Stokols and Shumaker (1981) suggest that the degree to which a particular setting satisfies the needs and goals of an individual determines his or her judgement of its quality. This quality judgement regulates the attachment to a place.

Riley (1992) states that attachments may not be to settings solely as physical entities, but may be primarily associated with the meanings of and experiences in a place, which often involve relationships with other people. Places are, therefore, contexts within which interpersonal, community and cultural relationships occur. So it is these social relationships, not just physical characteristics, to which people are attached (Altman and Low, 1992).

Giuliani and Feldman (1993) group the differences in the researchers' definitions of place attachment according to several characteristics:

- the content of the bond: affective, cognitive, and/or symbolic
- the valence of the bond: positive or negative
- the specificity of the bond.

Interest in understanding the attachments that people form with places can be found in a variety of disciplines. Scholars from diverse backgrounds such as family studies, psychology, geography, social ecology and gerontology have proposed various frameworks for understanding this phenomenon (Altman and Low, 1992). There have been many attempts to describe the process of interacting with an environment and the role of place attachment in determining spatial behaviours. Several models of people—place relationships have been put forward in an effort to provide a framework for how people develop ties to places, as identified in Inalhan and Finch (2004). Each framework places a different emphasis on the importance of stability and, therefore, on the consequences of leaving a place.

In recent years, place attachment has also gained increasing scientific interest in the field of natural resource management (Williams and Stewart, 1998). The importance of place attachment as a potential resource management tool stems from its non-economic perspective which can be used to understand the value of places. 'Building a better understanding of the values people attach to places could be a step toward a more integrated approach to resource management' (Warzecha and Lime, 2001). As suggested by Moore and Graefe (1994), 'an understanding of how recreationists perceive, choose, and relate to various settings is essential for researchers attempting to understand recreation behaviour and managers attempting to provide opportunities for satisfying recreation experiences'.

Some researchers choose to consider attachment as a broad concept, a super-ordinate category, whereby affects are designated as part of an entire system, such as a 'set of feelings'. Others seem to consider attachment a specific affect that is distinct from other kinds of affects which are part of the same system. Therefore we can identify in the literature the use of the terms attachment, territoriality and satisfaction with place interchangeably, without defining these concepts. There is a confusion regarding whether place attachment can be differentiated from other person/place terms such as territoriality and satisfaction with place. Some researchers (Stokols and Altman, 1987) have theorised that the concepts are intertwined, possibly with attachment either subsuming or being subsumed by the other two.

Scannell and Gifford (2010) propose a three-dimensional framework of place attachment (Figure 11.1). It is useful to structure the various definitions in the literature to provide a comprehensive understanding of place attachment.

- The person dimension (individual-cultural/group): Who is attached? To what extent is the attachment based on individually and collectively held meanings?
- The place dimension (the object of attachment including place characteristics: physical-social): What is the attachment to, and what is the nature of this place?
- The psychological process dimension (affect-cognition-behaviour): How are affect, cognition and behaviour manifested in the attachment?

Place attachment can be thought of as both a product/outcome (i.e. feeling attached) and a process (i.e. reasons for attachment) (Giuliani, 2002). As a product, place attachment is an emotional bond with a specific place. It is the experience of feeling attached and belonging to a place that can be stated at a point in time. This experience is multi-dimensional. As a process, place attachment is the appropriation of space via involvement with the local area. It is a continuous, dynamic process (Inalhan and Finch, 2004).

11.5 THE PROCESS OF PLACE ATTACHMENT

In everyday life, people are typically unaware of the existence of the bonds to certain places: it becomes apparent only in times of loss or in the idea of possible separation from the place. Thus, a relocation of an individual within an office, or a department to a new building, may provoke responses which remain largely subliminal.

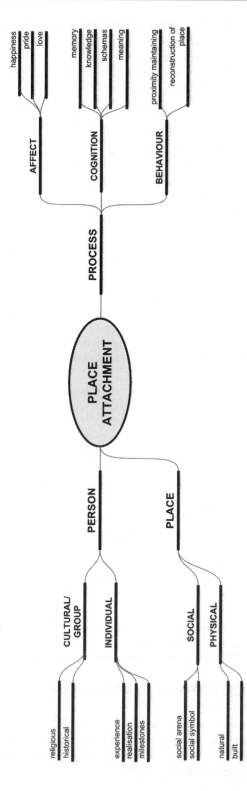

Figure 11.1 The tripartite model of place attachment. Reprinted from Scannell, L. and Gifford, R. (2010). Defining place attachment: A tripartite organizing framework. *Journal of Environmental Psychology, 30,* 1–10, with permission from Elsevier.

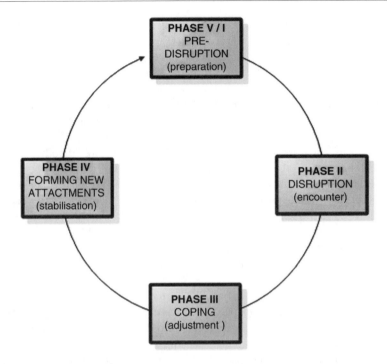

Figure 11.2 The transition cycle (Nicholson, 1990). Reproduced by permision of Professor Nigel Nicholson.

Kübler-Ross's five stages of grief model (1969), also known as the 'grief cycle', was developed initially as a model for helping dying patients to cope with death and bereavement; however, the concept also provides insight and guidance for coming to terms with personal trauma and change such as that caused by work redundancy and enforced relocation, and for helping others with emotional adjustment and coping.

Place attachments develop slowly but can be disrupted quickly and can create the need for a long-term phase of dealing with the loss and repairing or re-creating attachments to people and places.

There are discernible patterns across the phases of attachment and disruption. Place attachments are continuous and form a dynamic model of people/place bonds (Figure 11.2). After the development of secure place attachments (Phase 1), the loss of normal attachments creates a stressful period of disruption (Phase 2). Employee attempts at recovering what has been lost may take the form of holding on to work equipment, resisting a move to a new office location, lunching only with former colleagues, filing grievances and other actions to stop change, and other forms of written and vocal protest (Jefferys, 1985). It is followed by a post-disruption phase of coping with lost attachments (Phase 3). During this phase, the period of grief and despair for employees who survive an organisational transition is a necessary and valuable phase of grief. The grief work enables them to let go of the old and begin to attach to the new. This is the critical period for organisational help; it is important to provide time for grieving and support programmes (Jefferys, 1985). It is followed by the creation of new attachments (Phase 4). During this reorganisation phase, employees begin to look ahead to a new picture of 'Who am I?' at the other end of the transition. This is the next part of their healing-through-grief journey, and it requires a new identity. It is a time for identifying skills, developing new skills and even setting new work and career objectives (Jefferys, 1985) (Figure 11.2).

These four phases are interdependent, as qualities of the initial attachment or disruption can ease or exacerbate the stress of loss and difficulty of recreating attachments. Much of the challenge facing those with disruptions of place attachment is to negotiate reconciliation between the past (what has been lost) and the future. Certain aspects of pre-disruption attachment may forecast the extent and severity of the disruption and the availability and effectiveness of coping mechanisms (Brown and Perkins, 1992). The most stabilised conditions contain the possibility of future change, and therefore embody varying states of readiness for the onset of a new transition cycle (Nicholson, 1990).

Bowlby (1969) proposes that our first attachments provide a template or schema, or a set of expectations, that allow us to build other attachments later in life. He called this template the 'internal working model.' According to Brown and Perkins (1992), place attachments serve two basic functions: identity definition (e.g. autonomous selves and affiliated selves) and self-continuity/change (e.g. self-adaptation to new places) (Kleine and Baker, 2004). It has consequences both for short-term survival and in the longer term forming templates for later relationships.

In a workplace a personal bonding typically occurs when colleagues work side-by-side day after day. Relationships develop, people learn about one another and gain perspective of each others personal interests. A bond is a close personal relationship that forms between people working toward shared goals using collaborative efforts. Businesses often benefit from bonding in the office space. Employees who have bonded often work better together and employees who work well together improve productivity. Therefore there is a need to find a way to cultivate the team bonding experience in an office environment such as by designing workplace layout or changing the furniture to facilitate group gatherings. Hirschi (1969) argues a person follows the norms because they have a bond to society. The bond consists of four positively correlated factors: commitment, attachment, belief and involvement. Affective commitment refers to the employee's emotional attachment to, identification with, and involvement in the organisation. Employees with a strong selective commitment continue employment with the organisation because they want to do so. When any of these bonds are weakened or broken one is more likely to act in defiance.

Loss in the workplace is a product of change in the workplace, and this affects employees at all levels. When employees' bonds (with people, places, routines and even things) are broken, the physical and emotional pain of grief occurs. Therefore grief following relocation is attributed to loss of continuity of familiar surroundings and the existing social bonds (Fried, 1963). Although grief may be characterised by temporary disruption in business life, it is a generally an adaptive response (Jeffreys, 2005).

However, today, social theorists are often sceptical about the importance of place and place attachment, as people seem to be increasingly mobile in business life, and their social relations and other experiences become disembodied from physical location (Gustafson, 2001). Gustafson's (2001) findings in his study support the suggestions made by Mesch and Manor (1998) and Feldman (1990) that the geographical mobility of individuals does not necessarily contradict the importance and the role of place attachments. In other words, mobility does not necessarily preclude the continuity of place experiences (Feldman, 1990). Physical work environments (places) operate at two levels. On one hand they serve as the context of the work-related behaviours such as — an employee cannot sit unless there is a chair and desk for him. On the other hand, those environments influence people's behaviour such as when an employee cannot find a dedicated desk for himself; he occupies hot-desking areas all day long.

11.6 EVIDENCE OF PLACE ATTACHMENT AND TERRITORIALITY IN THE WORKPLACE

This section discusses the findings of work by the authors in relation to place attachment in the workplace. It has been shown that place disruptions (move experiences) interrupt the processes that bind people to their socio-environments. In order to understand the impact of this disruption, one must examine pre-existing conditions that influence the experience of attachments. It is also necessary to consider post-disruption conditions that influence how individuals cope with their losses and begin rebuilding ties to places and people. The difficulty of coping with loss and re-constructing place attachment is that individuals rarely appreciate the depth and extent of these attachments.

Relocation projects provide the opportunity for field experiments in which facilitators and inhibitors to the formation of place attachment can be identified during the whole process. As part of the British Institute of Facilities Management, Thames Valley Innovation Network (now defunct), the opportunity arose to study several work groups that were to experience office renovation within the same facility (Nationwide Headquarters, Swindon). The study was undertaken using a three-phase longitudinal approach conducted over one and a half years to monitor the process of place attachment (and detachment) starting two months before the relocation and ending 16 months after relocation.

In Inalhan's (2006) study, the Nationwide Property Services Department employees were studied in their natural work settings in order to evaluate the psychological impact of place attachment on employees during the process of change in the physical work environment. The phenomenon known as 'place attachment' (the emotional link formed by an individual to a physical environment) was considered in relation to its influence on people's response to the new environment and the move process. The phenomenon was assessed in terms of the meanings employees brought to their new work environment.

Based upon semi-structured interviews (Inalhan, 2006), a total of 51 interviews with employees of the case study organisation were conducted and analysed to monitor the move process. A hybrid of template analysis and Interpretative Phenomenological Analysis was used to analyse the qualitative data. The template analysis was applied in order to systematically code interview transcripts in which three key themes were identified to have the most central relevance to the aims of the research. These three themes are:

1. employees' reaction to change in the work environment
2. psychological processes and employees' behavioural changes
3. the aspects of employees' transactions with their environment.

Employee reaction — facilities managers who are responsible for spatial changes in the organisations often make the mistake of assuming that once a change is started, employees will see that it is going to take place and get on with it. However, any change, whether viewed positively or negatively, results in a certain degree of loss and an element of the unknown. It is often that fear of the unknown and the employees' losses that create the most anxiety regarding the change. In effect, employees go through emotions just like people coping with

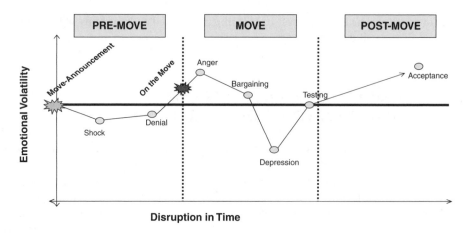

Figure 11.3 Emotional volatility versus disruption in time (adapted from Bourne and Bourne, 2002).

loss in their lives (Kübler-Ross, 1969; Jeffreys, 1995). Denial, anger, bargaining and acceptance are the emotional stages employees go through in order to cope with change. It is important to note that people may move back and forth in these phases and may overlap phases (Jeffreys, 1995).

Psychological impact — when the psychological impact on employees of introducing a change in the physical work environment was assessed after the announcement of the case study department's move, there were several recognisable patterned reactions (emotions) from employees towards the change in their physical environment. Denial, anger, bargaining and acceptance are the emotional stages employees go through in order to cope with change (Figure 11.3).

Employee transactions — during the preparation for the move, the move management team used a number of consultants to communicate and create the understanding of the need for change to their employees. This effective communication plan built employee trust and understanding of the project towards the readiness for change. From the moment the first announcements were made about the move project, although there was an initial nervousness about the uncertainty associated with the move, almost all employees were looking forward to this change. They were pleased that the business had recognised this need that they had long awaited.

However, employees' feelings changed during the move process as there were a number of delays between the announcement of the project and its actual implementation. This also affected employees' feelings towards being involved in the planning of the move. Specifically, they felt that the future was going to be no different from the present. For that reason the move project lost the momentous enthusiasm and buy-in from employees. This, in effect, installed general mistrust around the support for the project.

Once the reality set in, the employees showed signs of frustration and anger towards the project. They started to anticipate not only the implementation of the change but also the possible impacts of change on their work life. The ineffective communications and participation in the project played an important role in the perception of employees' loss of control. Most resistance in the case study department largely stemmed from their feeling of an inability to influence the changes that were taking place.

After the completion of the move project, the employees had a better understanding of the meaning of the change and were more willing to explore further and accept the inevitable new reality change brings about. However, at that stage, each employee's attitude produced a different response conditioned by feelings towards the change. While the move management team expected the successful outcome to prepare the ground for employees to adapt to the new situation, they found changing old habits and forming new ones particularly challenging.

11.6.1   Employees' predisposition to change

It is difficult to manage, encourage and support the practice of the new behaviours necessary to implement the desired change without understanding employees' old habits and attachments. To understand how these played a role in the case study department, the changes in behavioural patterns of the employees which occurred after the relocation were analysed and assessed individually and collectively.

As a basis for developing theoretical analysis, taxonomic terms are utilised based on transactional terms. In the employees' work, places are distinguished by whether they are occupied regularly/irregularly, whereas employees' association with their particular place/ lack of it (e.g. place-specific, which employees called 'desk bound', who has assigned a specific desk to work, and place-nonspecific, that is the 'flexible worker' who has no assigned desk), and employees' perceived attachment to the work environment (e.g. place-dependent and place-independent).

Given the widespread variations in the definitions and functions ascribed to place attachment, the qualitative interviews suggested that place attachment is not a singular phenomenon but more a set of related phenomena. These phenomena include satisfaction, territoriality, sense of community and place identity. The plurality of emotional bonds with place was assessed in the case study department's employees' evaluations of place attachments.

11.6.2   Attitudes towards existing workspaces

When employees were asked before relocation about their evaluative feelings and opinions of their current workplace, the 'secure base' concept, which is fundamental to attachment theory, was raised. While most employees were dissatisfied with the existing workplace, which was perceived as being disorganised and inconvenient to employees, their satisfaction with their workplace illustrates a predisposition towards place dependence by employees. In a way employees see their desk as a base to cover. This base allows them to set themselves up for the day. Although they do not use their desk regularly, they are comfortable with the fact that they can leave their belongings securely and expect to find them where they were left. Desks serve as an identifiable 'home base' which employees call their own.

The two employee groups who were either desk bound and/or had a desk of their own were looking forward to the change. Flexi workers were pleased that the business had recognised this need that they had long waited for and saw the change as an opportunity to be seized. Desk-bound employees were satisfied that the new environment would be more organised and effective in terms of communication and equipment.

11.6.3   Retrospective views of the change

Since the relocation there have been signs of an ongoing reassessment of the new work environment. There have also been noticeable behavioural changes arising from the new workplace layout and the changed social context. Some of the previously desk-bound employees have been pushed to work on flexible working practices, which requires two days a week working from home. These employees are the most affected by the move by not only losing their desk, but also needing to establish new identity and routines for the way they work. As explained in the qualitative analysis those employees are the ones most affected by the move of the department since they lost their desk (their 'base') and their team space. It is found in the qualitative interviews that this group's previous attachments to their desk and team members influenced them negatively to adapt to the new situation.

The second group to show similar qualities to the flexible working employees within a team is the desk-bound employees. This group, relative to the other groups, had not been directly affected by the move as they kept their desks in the new work environment. However, they were indirectly affected by the move as they lost the 'centre' of their team, because the new space planning did not allow employees belonging to a team to sit together. It is observed in the qualitative interviews that this group's attachments were not disturbed like their flexible working counterparts; they tended to show a dysfunctional attachment to their work environment and felt uncomfortable using the work environment on a shared basis. However, there were also some employees who showed positive adaptations to place and were in the process of assessing their work identity to become a place-independent worker.

The flexible-working employees with self-contained jobs are the ones that showed lower attachments to place and the social relations the place signifies in the old work environment. This might be part of the reason for their job role, which does not require working within a team, but more working and processing the team's outcomes. Also flexible-working employees were not satisfied with the old work environment since it did not fit the way they wanted to use the work environment. They particularly felt alienated in that environment since they were not recognised as a group. However, these groups of employees showed the quickest adaptation to the new environment; they also immediately formed attachments to the new workplace. These employees chose a particular area in the work environment to work and identified themselves with that place, in effect a flexible working community was formed. This case is particularly important as it proves that place attachments can be negative when a dependence on that particular place is low; however, to establish adaptation to an environment these bonds must be positive.

When those factors affecting the strength of the employees' attachment bonds to their work environment was confronted, individual profile, social networks and congruence between needs/goals and available resources were found to be critical.

The explanation of the relationship among these factors is that they satisfy the particular needs of individuals. The greater the particular needs the environment provides, the more people depend on their environment in general. This psychological interpretation of the bond between people and places has direct implications for the impact of relocation. The quantitative survey, which measures two dimensions (social and psychological), provides descriptive and confirmatory support for the links made between theoretical concepts. The quantitative data gathered from three groups of employees were found to support the qualitative interviews.

There are two critical conclusions that can be drawn from the analysis of interviews with employees. First, employees play a critical role in the successful implementation of workplace change projects. However, in practice there is a common neglect of the employees' experience of change and the bonds formed by employees to their previous work environment. As the previous bonds are unrecognised, the consequences of disturbing these bonds — along with employees' reactions to change — are left unmanaged. This implies the need for a cohesive perspective that can explain what is actually happening during the process of change in order to achieve a comprehensive understanding of employees' requirements.

Second, the applied changes in the physical environment can succeed to the extent that employees in organisations themselves adapt to the physical changes. The field study has helped show how assessing the place attachment process employees go through provides a means to an end as adaptation occurs.

The experience of place disruptions is a psychological factor that appears to play a crucial role in determining the later processes that bind people to their socio-physical environments. Therefore, the depth and extent of the experience of place attachment are explored as mediated by the personal meanings during the department's organisational move.

As change is viewed as the challenge to let go of the old, one can question the necessity of people's attachments to start new beginnings. However, it is important to acknowledge the people's previous attachments as it provides the means (opportunities) in the process of change to reassess outmoded thinking or beliefs; in return it can provide the strong foundations for future accommodation changes to be built upon. People may become connected to places in a variety of ways and for various reasons. Behaviour, reactions to others and feelings about a particular place may be influenced by the strength or type of attachment people experience and express.

As pointed out by Shumaker and Taylor (1983), the concept of attachment is vulnerable to mixed interpretations as it comprises the dual forces of loving the place versus moving on. Just as a parent/child attachment can be interpreted as dependence versus independence, conformity versus a secure base for exploration, and stunted growth versus emotional security, the concept of place attachment can have opposing implications for different people (Shumaker and Taylor, 1983).

The concept of place attachment suggested in this study, rather than creating controversy by becoming entrenched in a dualism, serves a broader (holistic) framework to explain and provide understanding of people/place relationships. In workplace change, while the attachments that no longer work for people are broken, there is a need to build new connections which can support people through the transition. Although individuals may react differently to disruptions of their attachment, a number of principles on the effects of interference can help guide understanding and management of place disruptions in the work environment. Understanding how place attachments are disrupted can help expand understanding of place attachment, as well as generate strategies to re-establishing familiarity, repairing attachment to place and stabilising place identity through new bonds.

In the literature a wide range of tools are used to implement a change programme, all sharing common characteristics of employee participation, consultation and collaboration.

11.7   FINDINGS

As reviewed in Inalhan (2006), in the current perception of workplace change the majority of studies focused on the workplace by assessing the outcomes. The new perspective discussed in this research shares similarities with one of the studies targeted to assess outcomes — post-occupancy evaluation. This involves collecting data that describe an employee's evaluations of various aspects of their work environment and of their behaviour and performance for a before-and-after comparison (Preiser *et al.*, 1988; Becker *et al.*, 1991; Preiser, 1999). The difference in the new perspective depends on monitoring and evaluating the workplace change process in a progressive cycle starting from the existing work environment where employees established their bonds to where employees need to adapt and change. This approach is important because it reduces the likelihood of assessing the outcomes of the new work environment on a 'treating the symptom' level; rather it goes to the root of the problem where place attachment plays a key role (Figure 11.4).

In workplace change, while the attachments that no longer work for people are broken, there is a need for building new connections which can support people through the transition. Although individuals may react differently to disruptions of their attachment, a number of principles of the effects of interferences can help guide understanding and management of place disruptions in the work environment. Understanding how place attachments are disrupted can help expand the understanding of place attachment, as well as generate strategies to establish new bonds to the new designed spaces.

11.8   IMPLICATIONS

Estimates suggest that about 15 million people work in offices in the UK (PFM, 2005). Today, like the Nationwide Property Services Department's facilities managers, many organisations' managers face a dilemma between managing 'workspace' and 'workplace' in the situations of workspace transformation. On the one hand, they are expected to manage the workspace and respond to the organisational pressures by rational decision making and reducing costs. On the other, they need to manage the workplace by acknowledging the meanings employees attribute to their workspace and the emotional charges these places hold for them (Vischer, 2005). Unless the decisions made in managing the workspace consider the workplace issues, the design decisions will have unintended and unanticipated consequences.

When there is change in the work environment, employees experience loss and grieving. Although a new physical work environment promises greater comfort and functionality, it is not enough to satisfy people to give up what they have and what they are familiar with. Therefore, there is a need to understand employees and their behaviour in their work environment to help companies work on effective definitions of the space employees occupy and successful implementations of physical environment change projects. The applied changes in the physical environment can succeed to the extent that individuals in the organisation themselves adapt to change. Therefore, change management requires understanding of how individuals adapt. However, in understanding the employees' transition one must also consider the role played by previously established familiar routines (habits) by employees towards their workplaces. In other words, place attachments which are known as previously formed emotional bonds to the physical environment play this critical role in establishing routines in the place.

The bonds between people and place are important and may be achieved by supporting the processes that contribute to these bonds. Such bonds may not only improve relationships between people and place, but also provide a greater sense of community and a stronger identity of place.

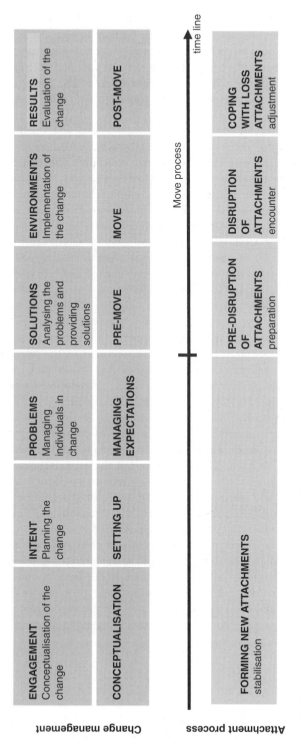

Figure 11.4   Place attachment in workplaces framework.

'Poorly designed offices could be costing British business up to £135 bn a year' was the headline in one of the facilities management publications (The Commission for Architecture and the Built Environment Report, 2005; Gensler Report, 2005; PFM, 2005). It is essential that organisations take a holistic view and integrate workplace management issues into their management system not only to get the full benefit from the workspace transformation but to strengthen and restore the terms of the 'socio-spatial contract' of employees. This would also prevent the damaging effects of failing to address employees' needs (Vischer, 2005).

It is also important for employers to attract and retain qualified and talented employees necessary for their success. Changing work attitudes among the younger generation have helped create a fluid workforce constantly able to search for that next perfect job. As employers strive to meet workforce challenges, they find it more difficult to differentiate themselves. At the same time, employees find the traditional attractors of job security, entitlements and benefits have been severely reduced or are non-existent (American Society of Interior Designers (ASID) Report, 2002). Facilities represent the organisation's identity. Understanding the influence of facility-related factors on the choices made by employees regarding a place to work is crucial (ASID Report, 2002). Therefore, developing a facility strategy to design a workplace where employees want to be is increasingly important. Creating a kind of workplace, a hallmark of learning organisations, is essential to attract and retain highly skilled employees. 'Design is not only the tangible artistic expression. It is a strategic investment that assists in formulating, translating and expressing an organisation's structure and style' (Becker, 1990). Place and people/place bonds can be integrated to the 'workplace of choice' strategies between management and employees that foster and support the skills, policies, processes and procedures needed to attract, develop and retain a diverse and committed workforce, who in turn create engaged customers.

In conclusion, the study of place attachment in the work environment is important for three reasons. First, it serves to foster a sense of community by supporting the integration of groups in an organisation. Second, it enables the attraction and retention of key staff. Third, it helps to identify and reflect the organisational culture.

Place attachment has a role to play in the environmental design of places. Facilities managers, designers and planners need to be aware that people become attached to places and that they are likely to encounter resistance from them. Relocation may be the 'psychological last straw' that prompts an employee to leave an organisation. In the knowledge that place attachment is a significant part of human well-being and psycho-cultural adaptation to an environment, the place attachment framework introduced in this study can be a valuable research tool since it can have potential application in predicting such behaviours related to the adjustment problems of employees whose environment has undergone change and to solve employees' problems of workspace design. Such solutions can also preserve or introduce design elements that nurture a sense of belonging in an organisation.

REFERENCES

Altman, I. and Low, S. (eds.) (1992). *Place Attachment*, Plenum, New York, NY.
ASID Report (2002). *Workplace Values: How Employees Want to Work*, sponsored by American Society of Interior Designers, Ecophon, Haworth and Vista Window Films.

ASTD (1990). American Society For Training And Development, Info-Line, October.

Becker, F.D. (1990). *The Total Workplace: Facilities Management and the Elastic Organization*, Van Nostrand Reinhold, New York, NY.

Becker, F., Quinn, K., Rappaport, A. and Sims, W. (1991). *New Working Practices*, IWSP Report, Cornell University, Ithaca, NY.

Briner, R.B. and Totterdell, P. (2002). The experience and management of emotion at work, in P. Warr (ed.), *Psychology at Work*, Penguin Books.

Bourne, M. and Bourne, P. (2002). *Successful Change Management in a Week*. Hodder & Stoughton, London.

Bowlby, J. (1969). Attachment, in Volume 1 of Attachment and Loss, Hogarth Press, London and Basic Books, New York (Penguin, Harmondsworth, 1971).

Briner, R.B. and Totterdell, P. (2002). The experience and management of emotion at work, in Peter Warr (ed.), *Psychology at Work*, Penguin Books.

Brown, B. and Perkins, D.D. (1992). Disruptions in place attachment, in Altman, I. and Low, S. (eds.), *Place Attachment*, Plenum, New York, NY.

Commission for Architecture in the Built Environment and British Council for Offices (2005). *The Impact of Office Design on Business Performance*. London.

Feldman R.M. (1990). Settlement identity: Psychological bonds with home places in a mobile society. *Environment and Behavior*, 22, 183–229.

Fisher, S. and Cooper, C. (eds.) (1990). *On the Move: The Psychology of Change and Transition*, Wiley, New York, NY.

Freeman, H.L. (ed.) (1984). *Mental Health and the Environment*, Churchill Livingstone, London.

Fried, M. (1963). Grieving for a lost home, in Duhl, L.J. (ed.), *The Urban Condition: People and Policy in the Metropolis*, Basic Books, New York, NY.

Fullilove, M.T. (1996). Psychiatric implications of displacement: contributions from the psychology of place. *American Journal of Psychiatry*, 153(12), 1516–23.

Gensler Report (2005). *These Four Walls the Real British Office*, Gensler Architecture, Design and Planning, London, available at: www.gensler.com/uploads/documents/TheseFourWalls_07_17_2008.pdf [Accessed 10 May 2005].

Giuliani, M.V. (2002). Place attachment, in Bonnes, M.; Lee, T, and Bonaiuto, M. (eds.) *Psychological Theories For Environmental Issues*, Ashgate, Aldershot.

Giuliani, M.V. and Feldman R. (1993). Place attachment in a development and cultural context. *Journal of Environmental Psychology*, 13, 267–274.

Grimshaw, R. and Garnett, D. (2000). Workplace democracy, in Nutt, B. and McLennan, P. (eds.), *Facility Management Risk and Opportunities*, Blackwell Science, Oxford.

Gustafson, P. (2001). Roots and routes: Exploring the relationship between place attachment and mobility. *Environment and Behavior*, 33(5), 667–686.

Hirschi, T. (1969). *Causes of Delinquency*, University of California Press, Berkeley and Los Angeles.

Inalhan, G. (2006). The role of place attachment on employees' resistance to change in workplace accommodation projects, unpublished PhD thesis, Reading University, Reading, UK.

Inalhan, G. (2009). Attachments: The unrecognised link between employees and their workplace (in change management projects). *Journal of Corporate Real Estate*, 11(1), 17–37.

Inalhan, G. and Finch, E. (2004). Place attachment and sense of belonging. *Facilities*, 22(5/6), 120–8.

Jeffreys, J.S. (1995). *Coping with Workplace Change: Dealing with Loss and Grief*, Crisp Learning.

Jeffreys, J.S. (2005). *Helping Grieving People: When Tears Are Not Enough: A Handbook for Care Providers*, Brunner-Routledge.

Kleine, S.S. and Baker, S.M. (2004). An integrative review of material possession attachment. *Academy of Marketing Science Review*, 1, 1–39.

Knoke, W. (1997). *Bold New World: The Essential Road Map to the Twenty-First Century*, Kodansha America.

Kübler-Ross, E. (1969). *On Death and Dying*, Simon & Schuster/Touchstone.

Mazumdar, S. (1992). Sir, please do not take away my cubicle: the phenomenon of environmental deprivation. *Environment and Behaviour*, 24(6), 691–722.

Mesch, G.S. and Manor, O. (1998). Social ties, environmental perception and local attachment. *Environment and Behaviour*, 30(4), 504–519.

Milligan, J.M. (1998). Interactional past and potential: the social construction of place attachment. *Symbolic Interaction*, 21(1), 1–33.

Milligan, J.M. (2003). Loss of site: organizational site moves as organizational deaths. *International Journal of Sociology & Social Policy*, 23(6/7), 115–52.

Moore, R.L and Graefe, A.R. (1994). Attachments to reactions settings: The case of railtrail users. *Leisure Sciences*, 16, 17–31.

Natarajan, S. (2003). Book Review: Richard Sennett: *The Corrosion of Character: The Personal Consequences of Work*, in the New Capitalism [online] Available from: http://www.stephanehaefliger.com/campus/biblio/017/17_82.pdf [Accessed 12 July 2010]

Nicholson, N. (1990). The transition cycle: causes, outcomes, processes and forms, in Fisher, S. and Cooper, C. (eds.), *On the Move: The Psychology of Change and Transition*, Wiley, New York, NY.

Nutt, B. and McLennan, P. (2000). *Facility Management Risk and Opportunities*, Blackwell Science, Oxford.

PFM (2005). A waste of space? *Premise Facilities Management*, available at: www.fmlink.com [Accessed 10 May 2005].

Preiser, W.F.E. (1999). Built environment evaluation: conceptual basis, benefits and uses, in Preiser, W.F.E. and Nasar, J.L. (eds.). *Directions in Person-environment Research and Practice*, Ashgate, Aldershot.

Preiser, W.F.E., Rabinowitz, H.Z. and White, E.T. (1988). *Post Occupancy Evaluation*, Van Nostrand Reinhold, New York, NY.

Riley, R.M. (1992). Attachment to the ordinary landscape, in I. Altman and S. Low (eds.), *Place Attachment*, Plenum, New York.

Scannell, L. and Gifford, R. (2010). Defining place attachment: A tripartite organizing framework. *Journal of Environmental Psychology*, 30, 1–10.

Sennett R. (1999). *Corrosion of Character: Personal Consequences of Work in the New Capitalism*, W.W. Norton Publications.

Shumaker, S.A. and Taylor, R.B. (1983). Toward a clarification of people-place relationships: A model of attachment to place, in Feimer, N.R. and Geller, E.S. (eds.), *Environmental Psychology – Directions and Perspectives*, Praeger, New York, NY.

Stegmeier, D. (2008). *Innovations in Office Design: The Critical Influence Approach to Effective Work Environments*, Wiley, Hoboken, NJ.

Stokols, D. and Altman, I. (eds.) (1987). *Handbook of Environmental Psychology*, Vols. 1 and 2, Wiley, New York, NY (reprinted by Krieger Publishing Company, Malabar, FL, 1991).

Stokols D. and Shumaker S.A. (1981). People In Places: A Transactional View Of Settings, in J. Harvey (ed.), *Cognition, Social Behaviour And Environment*, Hillsdale, NJ, Erlbaum.

Toffler, A. (1970). *Future Shock*, New York, Random House.

Topchik, G.S. (2001). *Managing Workplace Negativity*, AMACOM, New York, NY.

Vischer, J.C. (2005). *Space Meets Status: Designing Workplace Performance*, Routledge, London.

Warzecha C.A. and Lime, D.M. (2001). Place attachment in Canyonlands National Park: Visitors' assessments of setting attributes on the Colorado and Green Rivers. *Journal of Park and Recreation Administration*, 19(1), 59–78.

Williams D.R. and Stewart S.I. (1998). Sense of place: An elusive concept that is finding a home in ecosystem management. *Journal of Forestry*, 96(5).

Wood, L. and Giles-Corti, B. (2008), Is there a place for social capital in the psychology of health and place? *Journal of Environmental Psychology*, 28(2), 154–63.

12   Change Management and Cultural Heritage

Ana Pereira Roders and John Hudson

CHAPTER OVERVIEW

This chapter provides an introduction to the field of change management in the context of cultural heritage. It discusses cultural heritage as a motor for sustainable development and the role of a values-based management process for achieving this purpose. The concept of cultural significance is introduced, leading to an exploration of the differences between cultural heritage assessment and cultural heritage impact assessment. The chapter ends with conclusions and recommendations for facilities managers involved with change management in the context of cultural heritage assets.

Keywords: Change management; Cultural heritage; Cultural significance; Cultural heritage significance; Impact assessment.

12.1   INTRODUCTION

Cultural heritage is one of the greatest challenges for facilities managers involved in change management: to succeed, they need to reconcile the interests of future generations in the retention of the cultural significance of cultural heritage with the many other shorter term aims their own generation might perceive as vital.

This is not an impossible mission. Facilities managers need only to manage carefully the changes proposed by their generation and implement only the ones which can be shown not to endanger or irreversibly destroy the cultural significance of such cultural heritage assets. This process is not intended as a barrier to change; it is rather to enable change whilst simultaneously ensuring that the interests of future generations are taken into consideration.

Cultural significance plays a fundamental role in the change management of cultural heritage. However, to assess cultural significance is itself a challenge: this is especially so because the notion of what is culturally significant can change over generations and sometimes even between individuals. Cultural significance can be expected to change as a result of the developing history of cultural heritage, particularly as it is informed by new information and interpretation.

Facilities Change Management. Edited by Edward Finch.
© 2012 Blackwell Publishing Ltd. Published 2012 by Blackwell Publishing Ltd.

Therefore, the aim is to retain the cultural significance of cultural heritage by means of a multigenerational and multidisciplinary approach. The whole process to manage change, including the cultural significance assessment process and cultural heritage impact assessment process, needs to follow the same path. However, theories are easier to change than practices.

One reason may be that of professional tradition. Until very recently, architectural historians, urban planners, conservation specialists and related professionals dominated the decision-making process concerning the conservation of the built environment. In recent years, accountants, economists and financial analysts have been drawn more and more into this decision-making process, mainly due to rising costs and the increasing scarcity of funds (Throsby, 2003). Furthermore, community participation in the conservation of the built environment has also been drawing anthropologists, sociologists, psychologists and related professionals into the decision-making process. The growing environmental crisis, particularly the issue of global warming, has resulted in the involvement of ecologists, building physics engineers and related professionals to ensure that the conservation of the built environment reduces its environmental footprint.

The growing interdisciplinary nature of the subject has meant that a more holistic approach, based on sustainability, is slowly emerging, and providing a framework whereby the conservation of the built environment benefits from fitting these varied disciplines together. Facilities managers have potentially a key role in this situation, managing the built environment on several dimensions, mediating between the involved stakeholders, with their varied aims and needs.

12.2 CULTURAL HERITAGE

Cultural heritage may seem to be a somewhat arcane term of interest to conservation specialists, but in reality it represents a stage within the continuous process of evolution concerning the diversity of properties that are considered to be of cultural significance (UNESCO, 2005). These can include tangible heritage such as buildings, engineering structures, archaeological sites, historic areas or even whole landscapes, though these can also include intangible heritage assets such as customs, events and associations (UNESCO, 2003).

Within the built environment, a distinction can be drawn between designated and non-designated cultural assets. Designated assets, currently known as *cultural heritage*, are those that have been accorded special status and protection by the region, state or municipality. Non-designated assets are those that might have been recognised as having cultural significance but have not been accorded special status.

Most countries have legislation in place to designate cultural heritage assets, although the extent and consequences of designation vary from state to state. European states, for example, have signed up to the *Granada Convention for the Protection of the Architectural Heritage of Europe* (COE, 1985) which requires them to have systems for statutory designation in place. Moreover, the European Commission recently (9 March 2010) adopted a proposal to establish a European Heritage Label, with 64 sites listed to date (EC, 2010). Besides, 'strengthening European citizens' sense of belonging to the European Union, based on shared elements of history and heritage, as well as an appreciation of diversity, and to strengthen intercultural dialogue', the Label is seeking to 'enhance the value and profile of sites which have played a key role in the history and the building of the European Union' (EC, 2010).

Except for the intangible heritage entitled as Abolition of the Death Penalty, 1852/1867 (Portugal) currently listed as European Heritage, all other cultural heritage assets listed as European Heritage are tangible heritage. They range from movable (Raeren stoneware, German-speaking community) to immovable assets (Sacred Forests of Zemaitija and the Hill of Crosses, Lithuania). These latter often take the form of monuments, but there are also some archaeological sites, historic centres, industrial complexes and cultural landscapes listed.

Beyond Europe, there are also cultural heritage assets benefiting from worldwide protection. These encompass 731 cultural heritage assets (including the mixed assets) inscribed at the UNESCO World Heritage List together with another 180 natural heritage sites, protected under the *Convention Concerning the Protection of World Cultural and Natural Heritage* (UNESCO, 1972). This Convention is unique for its extensive adoption by, to date, 187 of the 192 UN countries (UNESCO, 2010).

There are close comparisons to be drawn between the conservation of cultural heritage and conservation of the natural environment and indeed the distinction between the two tends to blur on close examination. Navrud and Ready (2002) compared cultural heritage to environmental assets, due to their role as 'public goods'. Accordingly, they are both non-excludible, as it is 'technically infeasible to keep users from enjoying the good' and non-rival in consumption, as 'two different people can enjoy (consume)' them simultaneously, 'without interfering with each other's enjoyment'.

Against common expectations, the care of designated cultural heritage assets is perhaps easier to manage than non-designated assets. Normally they have their cultural significance already described in a Statement of Significance (or similar) produced by a governmental agency. In the UK, for example, cultural heritage assets are 'listed' for having special architectural or historic interest and have published descriptions of their particular important features. In order to carry out works that might affect the cultural significance of designated assets, the facilities manager will have to go through a statutory process in order to obtain permission for the development proposals. Permission may be subject to various, though established, conditions. In the UK, for listed buildings, this process is known as obtaining listed building consent. Whilst designation may restrict the type of work that can be carried out to a cultural heritage asset, the process is usually fairly transparent.

Non-designated assets may be more difficult to deal with. Traditional thinking, still present in most statutory legislation, tends to distinguish procedures according to the cultural significance of the built environment; from an extremely controlled process for designated cultural heritage assets with acknowledged cultural significance to a very permissive process for non-designated assets which are regarded as having little or no cultural significance.

More recent thinking depends less on the distinction between designated and non-designated assets and regards the man-made environment as being the outcome of a process of constant development and therefore having an overall cultural significance. An example of this would be historic landscape characterisation concept which is playing an increasingly important role in land use planning in the UK (Clark *et al.*, 2004). Individual sites, buildings and structures may contribute to the overall cultural significance of a place without necessarily being worth designating in themselves.

Particularly for facility managers dealing with the conservation of a built environment which hosts this variety of cultural significance (designated and non-designated assets), whilst still facing a great disparity between the processes (methods and criteria) decreed in the respective statutory legislation, this more recent thinking will certainly help them to decrease the level of complexity and accomplish their tasks more successfully.

12.2.1 Cultural significance

The term *cultural significance* came to prominence with the *Burra Charter* developed by Australia ICOMOS (International Council on Monuments and Sites) in 1979 (and revised in 1999), specifically developed to explicitly codify conservation principles in the Australian context, but which has later become influential in international conservation circles.

The Burra Charter has recognised that cultural significance was not limited to the physical fabric of a building or site but also extended to its setting, the way it was used, its contents and the knowledge that pertained to it. Accordingly, 'cultural significance is embodied in the place itself, its fabric, setting, use, associations, meanings, records, related places and related objects' (ICOMOS Australia, 1999). The reasons for regarding a cultural asset as important are often termed as *cultural values.*

Furthermore, the Burra Charter recognised that different groups and individuals would value the same cultural heritage asset in different ways and that cultural values would change over time. Cultural values are being qualified as extrinsic and subjective (Hodder, 2000), based on the changes of time and particular cultural, intellectual, historical and psychological frames of reference held by specific groups (Darvill, 1995). Moreover, different individuals or groups can attribute different and conflicting values (Ashworth, 1998). This rather contingent view of cultural values was in contrast to more traditional approaches that regarded cultural values as fixed and inherent in the assets themselves rather than constructed by those who used or contemplated them.

Both approaches have their share of truth. Research and practice has proven that different groups and individuals value the same cultural heritage asset in different ways and are influenced by their background knowledge. But, as previously suggested, the cultural values do not change in time (Pereira Roders, 2007). The so-called non-traditional cultural values such as the economic and political have always existed since the beginning of our civilisation. What changes is the importance (weighting) given to each one of the cultural values. So, when looking at the significance assessment results some cultural values appear to be excluded; but in practice, the assessors chose not to perform the respective surveys and consequently have these cultural values weighted as null. It does not mean that the assessed cultural heritage assets had no significance concerning those particular cultural values. Different ranges of cultural values have been developed and published over the past 20 years (Labadi, 2007). Its implementation remains quite loyal to the cultural values recognised at UNESCO's World Heritage Convention (1972, 2008). These are, respectively, the *historic, aesthetical/artistic, scientific* and *social* values.

A number of field experts have explored particular values and sets of values. Mason (2002) states that traditional practices of assessing cultural significance rely heavily on 'historical, art historical, and archaeological notions held by professionals, and they are applied basically through unidisciplinary means'. They have drawn attention to the neglect of *economic values,* 'a strong force shaping heritage and conservation'. Riganti and Nijkamp (2005) include the *political values* when defining cultural heritage as 'a set of recognised assets that reflect the historical, socioeconomic, political, scientific, artistic or educational importance of a good that has been created as a visible landmark by our ancestors'. As long ago as 1975 contributors at the Congress on the European Architectural Heritage, recognised within the Declaration of Amsterdam Council the *ecological values* of cultural heritage assets. Accordingly, 'the conservation of ancient buildings helps to economise resources and combat waste, one of the major preoccupations of present-day society' (EC, 1975). This anticipates later developments in thinking about sustainability in the context of cultural heritage. Age/historical values have long been defined by one of the pioneers in cultural significance and respective cultural values. Alois Riegl (1990 and 1903) defined two groups of cultural values: the memorial values (age, historical and intended memorial values) and the present-day values (use, art, newness and relative art values).

Recently, frameworks setting out a range of values have been developed. For example, in 2005, the EC published a set of methodological guidelines for cultural heritage assessments (Battaini-Dragoni, 2005). A list of categories to be considered in cultural heritage assessments included a broader range of cultural values such as historical, artistic/aesthetic, technological, religious/spiritual, symbolic/identity, scientific/research, social/civic, natural and economic values.

A recent survey of international documents published during the twentieth century recommending best practices on the conservation of the built environment have confirmed the reference to these eight cultural values — social, economic, political, historic, aesthetical, scientific, age and ecological values — when arguing the significance of cultural heritage assets (Pereira Roders, 2007). As expected, some cultural values were found to be far more widely referenced than others.

Facility managers could only benefit from such a multidimensional framework, as the decision-making process would be based on a broad range of assessments, addressing the significance of cultural heritage assets through the framework's varied dimensions. Moreover, they would also be able to determine patterns of weighting and relationships between the cultural values, either specifically per group of stakeholders or globally per generation.

An example follows to illustrate the difference between the cultural values used to justify the cultural significance of a cultural heritage asset and the broader range of cultural values. The Historic Centre of Guimarães (Portugal) was inscribed at the World Heritage List in 2001, under criteria (ii), (iii) and (iv). Criterion (ii) was proposed by the States Parties. Criteria (iii) and (iv) were later added by the World Heritage Committee, under the recommendation of its Advisory Body for cultural heritage assets, ICOMOS. Their justification revealed three natures of cultural values (UNESCO, 2001).

- Scientific values — stating that the outstanding universal value of Guimarães was reflected in its 'specialised building techniques' (criterion ii).
- Age values — as techniques employed at the site were found to have been developed 'in the Middle Ages' (criterion ii). Moreover, Guimarães also evidenced an 'evolution of particular building types from the medieval settlement to the present-day city (Figure 12.1), and particularly in the 15th–19th Centuries' (criterion iv).
- Historic values — as these techniques were found 'transmitted to Portuguese colonies in Africa and the New World' (criterion ii). Moreover, Guimarães is connected to 'the establishment of Portuguese national identity and the Portuguese language in the 12th Century' (criterion iii).

Figure 12.1 The Historic Centre of Guimarães, Portugal (© Tarrafa Silva 2010).

The range of cultural values attributed to Guimarães is much broader than the ones acknowledged on the justification of outstanding universal value. Already at the Nomination file (CMG, 2001), the States Party recognised certain qualities that helped assessors in identifying the following cultural values.

- Aesthetic values – wherein Guimarães was described as 'an ensemble and a testimony of urban development', with 'architectural characteristics (diversity of typologies, illustrating the evolution of the city at different periods) of outstanding universal value'.
- Ecological values – as Guimarães has, according to the States Parties, an harmonious 'integration with the landscape setting'.
- Economic values – in the 20th century, Guimarães 'expanded at an increasing speed owing to industrial development'. The historic centre remains 'intensely inhabited by the resident population'.
- Political values – as Guimarães was 'the feudal territory of the Portuguese Dukes who declared the independence of Portugal from León in the mid 12th Century'. Nowadays, 'the authenticity and the strong visual impact of the historic centre of Guimarães result from the unified protection strategies that have been implemented by the Municipal Technical Offices (GTL)' and that 'have been recognised by several international and national awards, such as the Prix Europa Nostra in 1985'.
- Social values – as the early history of Guimarães 'is closely associated with the forming of the national identity and the language of Portugal'.

Figure 12.2 illustrates the varied dimensions of cultural significance identified in the official documents produced during its nomination stage. There is no doubt that as some cultural values are found more frequently mentioned at the official documents, there are also differences in how relevant the facilities managers and respective decision-making process is based.

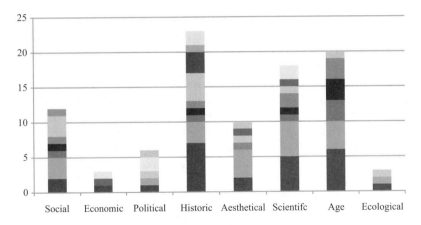

Figure 12.2 The cultural values of Guimarães (Tarrafa Silva and Pereira Roders, 2010).

12.3   CULTURAL HERITAGE MANAGEMENT

For the facilities manager charged with the care of cultural heritage assets the task of meeting modern user requirements whilst protecting their cultural significance can seem daunting. To the ones most active in the private sector, matters are often not helped by the cultural heritage professionals and statutory authorities who have their own practices that to outsiders may often seem unclear or inconsistent. To the ones most active in the public sector, matters do not get better. Whilst they are acquainted with the practices, the development and implementation of their management plans are often constrained by the statutory legislation and resources.

Historically, to conserve cultural heritage meant no more than to conserve it physically. However, recent decades have shown significant progress concerning the understanding of cultural heritage assets and respective conservation (Matero, 2003). Nowadays, conservation aims 'to maintain (and shape) the values embodied by the heritage – with physical interventions or treatment being one of many means towards that end' (Avrami *et al.*, 2000).

This implies that the management of cultural heritage also needs to evolve and accept the challenge, to make use of their cultural significance far beyond the traditional understanding and awareness, as an indispensable and recurrent milestone of their management process.

The Burra Charter, and consequently Australia, is seen as a pioneer state not only for having seeded the principles of values-based management for cultural heritage assets, but also for their implementation by the Statutory Agencies and respective property/facilities managers. Since 2004, the Australian Parliament enacted these principles in the legislation, establishing new statutory heritage lists, the National Heritage List and the Commonwealth Heritage List for Commonwealth-owned places (Altenburg, 2010). The Broken Hill, Australia, is only one of the many cultural heritage assets worldwide which are already profiting from the long-term benefits of following a values-based management. According to Altenburg (2010), such an approach has helped facilities managers 'to plan sustainably for heritage conservation'.

Central to the Burra Charter is the principle that when dealing with cultural heritage assets it is necessary to understand its cultural significance before undertaking any development project. The purpose is not to prevent change in the built environment but rather to ensure that any development project, either targeting the asset or its surroundings, is carried out in full knowledge of the impact it would have on the cultural heritage assets.

To insure that, it is crucial that the policies developed, to guide management and eventual changes, are fully based on the cultural significance of such assets. That is one of the advances contained in the Burra Charter. Rather than giving a static list of Do's and Don'ts, it described conservation as a dynamic process of change management, rather than as a static protectionist approach. Figure 12.3 details the values-based management process, from which the first three main stages are already to be found recommended in the Burra Charter:

1. understand significance
2. develop policy
3. manage in accordance with policy
4. change in accordance with policy.

> The concept of values-based management has implications for site managers and heritage professionals. Successful implementation requires management plans which actively involve site managers, a multi disciplinary team with a range of skills, practical and lateral thinking, flexibility and the ongoing commitment and involvement of the local community. Management plans should be living documents which inform management. (Altenburg, 2010)

The idea has been refined and developed into conservation plans and management plans (Kerr, 2000; Clark, 1999, 2001). In particular, these revisions introduced a further stage of assessing vulnerability into the process in order to explicitly identify threats to cultural significance. These ideas pervade much current conservation guidance. In England, for example, they are central to English Heritage's (2008) Conservation Principles: Policies and Guidelines.

12.3.1 Cultural heritage assessments

The Burra Charter and its later offshoots emphasise the importance of understanding a cultural heritage asset before carrying out work on it so that its cultural significance can be protected as far as possible in any conservation work. The development of such an understanding requires investigation into the cultural heritage asset on several dimensions.

The scale and complexity of such an investigation normally depends upon how much information is already in the public domain and upon the likely importance and complexity of the cultural heritage asset. For the more complex or important assets it may be necessary to engage the services of specialist consultants to undertake the necessary investigations.

Cultural heritage assessments can be seen as a development of a long tradition of scholarly evaluation and they may take many forms and serve a variety of purposes. Some are a part of the process used to identify and designate particular assets for legal protection. For example, in England there have been two major national surveys of historic buildings and structures using a standardised methodology and assessment criteria in order to

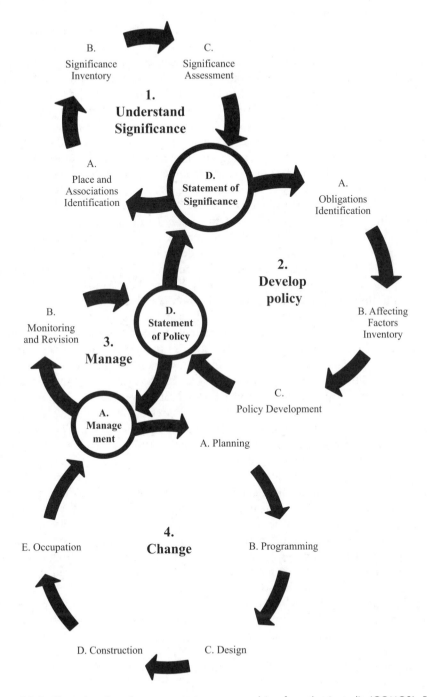

Figure 12.3 The values-based management process, evolving from the Australia ICOMOS's Burra Charter (ICOMOS Australia, 1999).

produce a uniform list for statutory protection (Robertson, 1993). There have been parallel developments in other European countries, such as France (Chatenet, 1995) and Germany (Wulf, 1997). Some national surveys such as the work of the Royal Commission on the Historical Monuments of England have been less focused on specific types of statutory designation (Sargent, 2001) whilst others such as Pevsner's guides (Architectural Guides and Buildings Books Trust) to the buildings of the British Isles were intended to bring an awareness of architecturally important buildings to the general public (Pevsner, 2010). Whilst cultural heritage assessments have been most commonly developed with respect to material artefacts, particularly buildings and archaeological sites, they can also be used for intangible heritage. Japan, for example, designates individuals or groups of people with particular skills in arts or crafts as 'Preservers of Important Intangible Cultural Properties' or, more colloquially, 'Living National Treasures' under the 1950 Law for the Protection of Cultural Properties.

Cultural heritage assessments involve both the description of assets and an evaluation of their cultural significance in terms of values, as discussed in the previous section. Investigation of a cultural heritage asset may require the use of a number of techniques depending upon the nature of the asset and the depth of analysis needed for the purpose of the assessment. For a comprehensive assessment an accurate recording of the asset is often required. For standing structures this may involve the use of traditional measured surveys but these might be augmented by photogrammetry or 3D laser scanning, particularly where parts of the asset are not easily accessible to the tape measure. Measured drawings should be accompanied by detailed observations and notes together with an adequate photographic record. The record provided by such surveys forms the base line against which future proposals for change can be assessed in terms of their impact upon cultural significance. A major survey can be time consuming and costly so it is important to specify an appropriate level of survey to the importance and complexity of the cultural heritage asset being considered (Bryan et al., 2009). In many cases there may also be documentary material including previous surveys or interpretations and archives relating to stages in the asset's history. These may also provide important evidence that can be used in the assessment.

Once the asset has been adequately described and recorded, the assessment enters the process of evaluation. This involves making judgements about the cultural significance of the asset against criteria developed from an appropriate set of cultural values, typically social, economic, political, historic, aesthetical, scientific, age and ecological values (Pereira Roders, 2007). Where the cultural heritage assessment is being used as part of a selection process to designate particular assets as worthy of statutory protection an exact set of criteria may be developed as a threshold for inclusion. Where selection is not involved, the appraisal may be more discursive and scholarly. The process often involves an attempt to distinguish between features of the asset that are of major importance and hence should be passed intact to future generations and those that are of lesser importance and hence can be the subject of negotiation in the change management process.

Community participation is an important part of the cultural heritage assessment process. Stakeholders in the asset have a legitimate right to involvement and this can take a variety of forms such as workshops, focus groups and surveys. Where individuals or groups are found to hold conflicting values over an asset a resolution process may have to be brought into play.

It should be emphasised that cultural heritage assessments are independent from any specific proposal for change to a cultural heritage asset. In some cases they may lead to a set of general policies for the management of an asset as in the conservation plan (Clark, 1999; Kerr, 2000). However, specific development proposals should be subject to a cultural heritage impact assessment.

12.3.2 Cultural heritage impact assessments

Proposals for change in the historic environment will have an impact upon the cultural significance of that environment. Once the significance of the historic environment has been researched and understood, the next stage is to assess the impact of the proposed changes. This is not necessarily a complicated task. As Clark (2001, p. 22) observes, 'Impact analysis is not a particularly special, unusual or complex process; it is simply a codification of the basic analysis undertaken by any competent conservation adviser.'

When a statutory authority makes a decision on whether or not to allow development proposals for work on a cultural heritage asset to take place it is making an impact assessment. The proposals are evaluated in terms of potential damage or benefits they may make to the significance of the asset; this analysis may also include a consideration of a wider set of social, environmental or economic benefits. In some cases alternative proposals may be compared to determine which has the least impact on cultural significance. For small-scale interventions this process is often informal and based on the professional judgement of the individuals concerned. For larger-scale interventions a more formal method of analysis may be adopted.

Cultural heritage impact assessments form part of the wider group of analytic approaches for evaluating the impacts of development that include Environmental Impact Assessment (EIA) and Social Impact Assessment (SIA): all these techniques adopt the same broad methodological approach. EIA, in particular, is a long established approach that has been widely adopted as part of the land-use planning system in many countries (Glasson *et al.*, 2005; Morris and Therivel, 2008).

European states are required to adopt impact analysis in their land-use planning systems by the Directives on Environmental Assessment — Directive 85/337/EEC (EC, 1985), amended by Directive 97/11/EC (EC, 1997) and the Directive on Strategic Environmental Assessment — Directive 2001/42/EC (EC, 2001). EIA is mandatory for certain types of development such as major transport infrastructure projects and may be required for a broader range of developments, depending on the enactment in a particular state.

It is useful here to distinguish between cultural heritage as a factor to be evaluated as part of a general EIA and cultural heritage impact assessment (CHIA) in particular. EIA usually takes as its focus a major development project such a road, industrial plant or airport that may affect a number of cultural heritage assets or indeed whole areas of cultural significance. The analysis will often include cultural heritage as one factor to be evaluated amongst a range of other socio-economic and bio-physical factors when considering a development proposal. These factors may be given a weighting in accordance with their perceived importance in the process. A number of commentators have discussed weaknesses in the evaluation of cultural heritage contained within current approaches to EIA (Teller and Bond, 2002; Bond *et al.*, 2004; Dupagne *et al.*, 2004; Jones and Slinn, 2008).

The SUIT method (Figure 12.4), defined to assist sustainable development of urban historical areas, was developed to ensure that EIA and SEA procedures 'are used to steer the parallel (re)definition of both the long-term active conservation strategies and the proposed

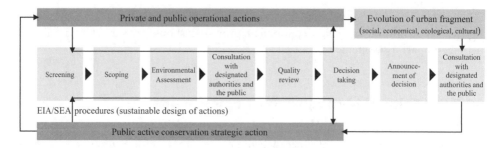

Figure 12.4 The SUIT method (adapted from Dupagne *et al.*, 2004).

plan, programme or project, in order to make them converge towards a win-win solution' (Dupagne *et al.*, 2004).

CHIA, in contrast to EIA, is more limited in scope and focused specifically on proposals for change to a particular asset or area of cultural significance and the analysis is confined to the impacts on cultural significance.

Both EIA and CHIA adopt a broadly similar approach. Once the overall scope of the study has been defined a baseline survey is carried out to provide a reference point against which impacts can be measured. In a CHIA this would be the assessment of cultural significance as discussed in the previous section. The survey might include both a desktop study and field research. Once significance has been established, the impact of the proposals is assessed. Impacts may be positive or negative.

A positive impact, for example, might be to provide better access and understanding of a site to the general public. A negative impact might be the loss of important historic features. Impacts can also be direct or indirect. Direct impacts are those that have an immediate effect on the site, such as loss or additions to the historic fabric. Indirect impacts are those that may not result in immediate change to the physical fabric but may have wider implications (e.g. increase in visitor numbers or changes to the wider visual setting of the site). Where there are alternative proposals for the site these can be evaluated to determine their relative impacts. In EIA the outcome of the process of evaluating likely impacts may be formally presented as an environmental impact statement (EIS).

Even though there might be cases where cultural heritage assessments and cultural heritage impact assessments seem similar, it is important that each development project proposal has its impact assessed independently for the affected cultural heritage assets. The relative importance of development proposals to the cultural significance of an asset may vary considerably from case to case and generalising impact assessment results could lead decision makers into erroneous conclusions and, consequently, irreversibly compromise the cultural heritage assets in question.

In some circumstances the wider importance of the development proposals that have negative impacts may be deemed to outweigh the cultural significance of the asset. For example, the addition of the second runway to Manchester Airport in the UK resulted in the demolition of a number of cultural heritage assets (Griggs *et al.*, 1998; Butcher, 2010). Such decisions are controversial and may result in major objections from communities and amenity groups. In other circumstances the negative impacts on the cultural significance of the site may considerably outweigh the potential benefits of a development proposal and the development is therefore terminated.

In most cases there is likely to be a process of mitigation, i.e. attempting to reduce negative impacts and increase beneficial impacts. Mitigation may take many forms. Where some loss of cultural heritage assets is inevitable there may be a requirement to compile detailed records before the development takes place; this is sometimes referred to as preservation by record. Although it is usually desirable from a conservation perspective to retain a cultural heritage asset in its original use and location where possible, it is often recognised that this may no longer be feasible: modifications to the built environment may be required to allow new uses to be found. Mitigation has an important role in feedback into the design of proposals. It is often possible through careful design to minimise negative impact through matters such as the careful choice of materials, avoidance of existing fabric of major cultural significance, and location of additional structures and services. General conservation principles that can be used to guide sensitive design in the context of the historic environment are well established and widely available (English Heritage, 2008).

12.4 CHANGE MANAGEMENT AND CULTURAL HERITAGE

The change management of cultural heritage assets has proven to be a very dynamic practice, with the constant evolution of notions, methods, access to information, etc. Therefore, it is fundamental for the facility managers involved with change management in the context of cultural heritage assets to keep abreast of such changes and evolve their practices in tandem.

A major attitudinal change has occurred over the last decades and greater attention needs to be paid to the specific attributes and elements composing these cultural heritage assets that illustrate their cultural significance, authenticity and integrity. By having them periodically monitored, facilities managers gain more freedom, as it is clear what is happening and why specific development proposals need to be declined.

Cultural significance is becoming more and more multidimensional: the measurement tools and composition of expert teams also need to become more multidimensional. Facilities managers can play a fundamental role on this matter and ensure that their advisory bodies are impartial. Whenever their own teams have an inadequate spread of expertise, they should be complemented with external consultancy on the lacking fields.

Values-based management is one practice that has proven to be very successful in terms of assisting facilities managers involved with change management in the context of cultural heritage assets. More important than the chosen practices, however, is the clarity concerning the cultural values supporting the decision-making process. This in turn should be related to the respective impact (positive and negative) of the proposed changes on the cultural significance of the cultural heritage assets under their guardianship. 'The heritage of the past is the seed that brings forth the harvest of the future' (Fraser, 1933–1935). Thus, it is up to the facilities managers to ensure that their decision making does not compromise future generations who are likely to benefit from these assets of outstanding cultural significance.

REFERENCES

Altenburg, K. (2010). Values based management at cultural heritage sites, in Amoêda, R., Lira, S. and Pinheiro, C. (eds.), *Heritage 2010, Heritage and Sustainable Development.* Greenlines Institute for the Sustainable Development, Barcelos.

Ashworth, G. (1998). Heritage, identity and interpreting a European sense of place, in Uzzell, D. and Ballantyne, R. (eds.), *Contemporary Issues in Heritage and Environmental Interpretation: Problems and Prospects*, pp. 112–32. The Stationery Office, London.

Avrami, E., Mason, R. and De La Torre, M. (eds.) (2000). *Values and Heritage Conservation*, Getty Conservation Institute, Los Angeles.

Battaini-Dragoni, G. (ed.) (2005). *Guidance on Heritage Assessment*, Council of Europe, Strasbourg.

Bond, A. *et al.* (2004). Cultural heritage: Dealing with the cultural heritage aspect of environmental impact assessment in Europe. *Impact Assessment and Project Appraisal*, 22(1), 37–45.

Bryan, P., Blake, B. and Bedford, J. (2009). *Metric Survey Specifications for Cultural Heritage*, English Heritage, Swindon.

Butcher (2010) 'Aviation: Manchester's second runway, 1993–2001', House of Commons Library Standard Note: SN/BT/101 Online: www.parliament.uk/briefingpapers/commons/lib/research/briefings/snl (accessed 7th July 2010).

Chatenet, M. (1995). The inventory and protection of the heritage in France. *Transactions of the Ancient Monuments Society*, 39, 41–50.

Clark, J., Darlington, J. and Fairclough, G. (eds.) (2004). *Using Historic Landscape Characterisation*, English Heritage and Lancashire County Council.

Clark, K. (1999). *Conservation Plans in Action*, English Heritage, London.

Clark, K. (2001). *Informed Conservation*, English Heritage, London.

CMG (2001). *Historic Centre of Guimarães: Nomination file* (available online http://whc.unesco.org/en/list/1031/documents/ [accessed on 01/07/2010]).

COE (1985). *Granada Convention for the Protection of the Architectural Heritage of Europe. European, Treaty Series No.121*. Strasbourg, Council of Europe (COE).

Darvill, T. (1995). Value systems in archaeology, in Cooper, M., Firth, A., Carman, J. and Wheatley, D. (eds.), *Managing Archaeology*, pp. 40–5, Routledge, London.

Dupagne, A., Ruelle, C. and Teller, J. (eds.) (2004). *Sustainable Development of Urban Historical Areas through an Active Integration within Towns*, Research Report no. 16. European Commission (EC), Brussels.

EC (1975). *The Declaration of Amsterdam*, Congress on the European Architectural Heritage (available online http://www.icomos.org/docs/amsterdam.html [accessed on 01/07/2010]).

EC (1985). *Directive 85/337/EEC of 27 June 1985 on the assessment of the effects of certain public and private projects on the environment* (available online http://ec.europa.eu/environment/eia/full-legal-text/85337.htm [accessed on 01/07/2010]).

EC (1997). *Directive 97/11/EC of 3 March 1997 amending Directive 85/337/EEC on the assessment of the effects of certain public and private projects on the environment* (available online http://ec.europa.eu/environment/eia/full-legal-text/9711.htm [accessed on 01/07/2010]).

EC (2001). *Directive 2001/42/EC of 27 June 2001 on the assessment of the effects of certain plans and programmes on the environment* (available online http://eur-lex.europa.eu/LexUriServ/LexUriServ.do?uri=CELEX:32001L0042:EN:NOT [accessed on 01/07/2010]).

EC (2010). *European Heritage Label* (available online http://ec.europa.eu/culture/our-programmes-and-actions/doc2519_en.htm [accessed on 01/07/2010]).

English Heritage (2008). *Conservation Principles: Policies and Guidelines*, English Heritage, London (available online at http: //www.english-heritage.org.uk/publications/conservation-principles-sustainable-management-historic-environment/conservationprinciplespoliciesguidanceapr08web.pdf/ [accessed on 21/06/2010]).

Fraser, J.E. (1933–1935). *Inscription on the 'Heritage' Statue*, U.S. National Archives Building, Washington, D.C.

Glasson, J., Therivel, R. and Chadwick, A. (2005). *Introduction to Environmental Impact Assessment*. Routledge, Oxon.

Griggs, S., Howarth, D. and Jacobs, B. (1998). Second Runway at Manchester. *Parliamentary Affairs*, 51(3), 370–369.

Hodder, I. (2000). Symbolism, meaning and context, in J. Thomas (ed.), *Interpretive Archaeology: A Reader*, pp. 86–96, Leicester University Press, London.

ICOMOS Australia (1999). *The Australia ICOMOS Charter for the Conservation of Places of Cultural Significance* (available online http://australia.icomos.org/wp-content/uploads/BURRA_CHARTER.pdf [accessed on 29/06/2010]).

Jones, C.E. and Slinn, P. (2008). Cultural heritage in EIA – Reflections on practice in North West Europe. *Journal of Environmental Assessment Policy and Management*, 10(3), 215–238.

Kerr, J.S. (2000). *Conservation Plan* (5th edn.), The National Trust of Australia, Sydney.

Labadi, S. (2007). Representations of the nation and cultural diversity in discourses on World Heritage. *Journal of Social Archaeology*, 7(2), 147–170.

Matero, F. (2003). 'Preface', in Teutonico, J.M. and Matero, F. (eds.), *Proceedings of the Forth Annual US/ ICOMOS International Symposium on Managing Change: Sustainable*, April 2001, Getty Conservation Institute, Los Angeles.

Mason, R. (2002). Assessing values in conservation planning: Methodological issues and choices, in M. de la Torre (ed.), *Assessing the Values of Cultural Heritage. Research Report*, pp. 5–30, The Getty Conservation Institute, Los Angeles.

Morris, P. and Therivel, R. (eds.) (2008). *Methods of Environmental Impact Assessment*, Spon Press, London.

Navrud, S. and Ready, R.C. (2002). *Valuing Cultural Heritage: Applying Environmental Valuation Techniques to Historic Buildings, Monuments and Artefacts*, Edward Elgar, Cheltenham.

Pereira Roders, A. (2007). *RE-ARCHITECTURE: Lifespan Rehabilitation of Built Heritage*, Eindhoven University of Technology, Eindhoven.

Pevsner, N. (2010). *Pevsner Architectural Guides and Buildings Books Trust* (available online at http://www. pevsner.co.uk/ [accessed 5 August 2010]).

Riegl, A. and Scarrocchia, S. (1990). *Il culto moderno dei monumenti, il suo carattere e i suoi inizi*, Nuova Alfa Editoriale, Bologna, translation of Riegl, A. (1903). *Der Moderne Denkmalkultus, SeinWesen und seine Entstehung*, Verlage von W. Braumüller, Vienna, (in Italian).

Riganti, P. and Nijkamp, P. (2005). Benefit transfers of cultural heritage values: How far can we go? *Proceedings of the European Regional Science Association (ERSA) Conference*, European Regional Science Association, Amsterdam.

Robertson, M. (ed.) (1993). Listed buildings: The National Resurvey of England. *Transactions of the Ancient Monuments Society*, 37, 21–94.

Sargent, A. (2001). RCHME 1908–1998: A history of the Royal Commission on the Historical Monuments of England. *Transactions of the Ancient Monuments Society*, 45, 57–80.

Tarrafa Silva, A. and Pereira Roders, A. (2010). The cultural significance of World Heritage cities: Portugal as case study, in Amoêda, R., Lira, S. and Pinheiro, C. (eds.), *Heritage 2010, Heritage and Sustainable Development*, Greenlines Institute for the Sustainable Development, Barcelos.

Teller, J. and Bond, A. (2002). Review of present European environmental policies and legislation involving cultural heritage. *Environmental Impact Assessment Review*, 22(6), 611–632.

Throsby, D. (2003). Determining the value of cultural goods: How much (or how little) does contingent valuation tell us? *Journal of Cultural Economics*, 27, 275–285.

UNESCO (1972). *Convention concerning the protection of the world cultural and natural heritage* (available online http://whc.unesco.org/archive/convention-en.pdf [accessed on 29/06/2010]).

UNESCO (2001). *Decision 25COM X.A - Historic Centre of Guimarães (Portugal)* (available online http:// whc.unesco.org/en/decisions/2303 [accessed on 29/06/2010]).

UNESCO (2003) *The Convention for the Safeguarding of the Intangible Cultural Heritage* (available online http://www.unesco.org/culture/ich/doc/src/01852-EN.pdf [accessed on 29/06/2010]).

UNESCO (2005). *Convention on the Protection and Promotion of the Diversity of Cultural Expressions* (available online http://unesdoc.unesco.org/images/0014/001429/142919e.pdf [accessed on 29/06/ 2010]).

UNESCO (2008). *Operational Guidelines for the Implementation of the World Heritage Convention* (available online http://whc.Unesco.org/archive/opguide08-en.pdf [accessed on 29/06/2010]).

UNESCO (2010). *States Parties* (available online http://whc.unesco.org/en/statesparties/ [accessed on 29/ 06/2010]).

Wulf, W. (1997). German inventory and heritage - A fateful genesis from history, politics and science. *Transactions of the Ancient Monuments Society*, 41, 45–58.

Index